PHILOSOPHY OF SCIENCE AND ITS DISCONTENTS

THE CONDUCT OF SCIENCE SERIES

Steve Fuller, Ph.D., Editor
Center for the Study of Science in Society
Virginia Polytechnic Institute

Philosophy of Science and Its Discontents, Second Edition
Steve Fuller

The Scientific Attitude, Second Edition
Frederick Grinell

Politics and Technology
John Street

Philosophy of Science and Its Discontents

Second Edition

Steve Fuller

THE GUILFORD PRESS

New York   London

© 1993 The Guilford Press
A Division of Guilford Publications, Inc.
72 Spring Street, New York, NY 10012

Printed in the United States of America

This book is printed on acid-free paper.

Last digit is print number: 9 8 7 6 5 4 3 2

Library of Congress Cataloging-in-Publication Data

Fuller, Steve, 1959–
  Philosophy of science and its discontents / Steve Fuller. —2nd ed.
    p. cm. — (Conduct of science series)
  Includes bibliographical references and index.
  ISBN 0-89862-020-1 (alk. paper)
  1. Science—Philosophy. 2. Science—Methodology. 3. Knowledge, Theory of. I. Title. II. Series.
Q175.F925 1993
501—dc20   92-38514

Contents

Acknowledgments

My interest in reviving the normative dimension in the philosophy of science has motivated me to work closely over the last five years with cognitive and social psychologists. It started when Marc De Mey helped me organize the 1987 Sociology of Science Yearbook conference at the University of Colorado. He put me on to Barry Gholson's Metascience group at Memphis State University, where I participated in an intense and productive weeklong seminar in July 1988 on the mutual relevance of psychology of science and social epistemology. Of that group, Art Houts and Will Shadish have been especially helpful in sharpening my understanding of the potential for an experimental psychology of science. Shadish and I have since developed a research agenda for the social psychology of science, which pools the intellectual resources of some of the best social psychologists currently writing (Shadish and Fuller 1992). Among the ones I originally met at the Yearbook conference were Donald Campbell and Mike Gorman, who represent the best of the old and the new generation of psychologists in their perspicacity of their research into human beings.

Philosophy and sociology of science continue to be the main stimulants to my thought. In England, the annual conference on Realism in the Human Sciences, led by Roy Bhaskar and William Outhwaite, has caused me to rethink my views on scientific realism. In the United States, similar proddings have come from Richard Boyd and Ron Giere. Among the sociologists, Bruno Latour and Steve Woolgar never fail to arouse my intellectual passions. More recent encounters with Joe Rouse and Paul Roth continue to convince me that the space bounded by continental social philosophy, analytic philosophy of mind, and philosophy of social science is where the action is in our discipline today. Both have helped shaped the way I drew "my map of the field," which appears for the first time in this edition of the book.

Along with these influences, I must count my graduate students. Those who attended my social epistemology seminars at the University of Colorado were particularly helpful in refining my views on the relation of analytic epistemology to cognitive science that I have had no reason to change from the first edition: Brian Beakley, Chris McClellan, Franz-Peter Griesmaier, David Mertz, Paul Saalbach, and Charles Wallis. On matters of politics and social science, I thank Sandra Gudmundsen, who has kept me close to Habermas and has also gently nudged me in the direction of feminism. If for employment purposes only, none of these students should be held accountable for the views expressed in this book. At Virginia Tech, I have been blessed with an equally engaging crop of students, who provided much local intellectual sustenance during an incredibly turbulent period at the Science Studies Center, when this new edition was written: Ranjan Chaudhuri, Jim Collier, Garrit Curfs, and especially Bill Lynch receive my deepest gratitude. Only my wife, Sujatha Raman, has been a more lasting source of support.

Three people spent much of the summer of 1988 reading the original manuscript draft of the first edition, offering very helpful and detailed remarks. They convinced me to develop half of it and table the other half for future works (much of which appears in Fuller 1992a). Given their diverse backgrounds and interests, their convergence on what was good, bad, and ugly about the manuscript was gratifying. Many thanks to Brian Baigrie, Tom Nickles, and Steve Turner.

I hope this second edition will be of use to students or anyone interested in understanding the very latest developments in the philosophy of science and neighboring fields. I owe a debt of gratitude to Seymour Weingarten and Peter Wissoker at Guilford Press for agreeing to put this version out in paperback. In the past three years, the first edition has received its share of plaudits and criticisms. I especially thank Ron Giere, Peter Dear, Harold Brown, Jonathan Adler, Rex Welshon, and Mike Malone, for their searching criticisms, an omnibus response to which I try to provide at the end of this edition.

Finally, once again, this book is dedicated to the person who has had to wrestle with the ideas in this book almost as long as I have, who stood with me during some very strange times— Professor Stephen Downes of the University of Utah.

Steve Fuller

Introduction

To those who look at philosophy from the outside, philosophers of science look the most like philosophers. However much they vary in other respects, philosophers of science share an interest in determining how human beings have managed to produce as much knowledge as they have in the relatively short period of time that they have been producing it. This shared interest rests on two assumptions. First, it assumes that there is a prima facie discernible pattern to the part of human history pertaining to the production of knowledge. Second, it assumes that there is something significant about knowledge as a product of human endeavor that warrants serious study. Explaining and defending these two assumptions has been the subject of wide-ranging debate that has struck most onlookers as quintessentially philosophical. After all, who else spends their time so unabashedly extracting patterns from history and divining their significance?

In my first book, *Social Epistemology*, I attempted to reclaim the classical mission of the philosophy of science, few traces of which can be found in the field today. Once Kant legitimized the study of knowledge as independent from the study of ultimate reality, it was no longer necessary to answer all the skeptic's questions before proceeding with the epistemic enterprise. The nineteenth century grasped this point very well and embodied it most successfully in the positivist movement, whose main philosophers were Auguste Comte and John Stuart Mill. Comte and Mill took the problem of knowledge to be a largely normative one: What is the most valuable form of knowledge? How can we get more of that form of knowledge? Would it make sense to have only that form and no other? Answers to these questions turned on the social organization of inquiry, understood both externally (science in the society at large) and internally (science as a society in small). As science became more a society in small, the positivists and their fellow travelers began to lose touch with science in the society at large. This is the period in which we still live. The leading players in twentieth century philosophy of science—the logical positivists, the

Popperians, the Kuhnians—have managed to retain the scope of the nineteenth century enterprise, albeit more abstractly, and with the social dimension relegated to a few telling metaphors, as in the "revolutions" that occur on the pages of Kuhn's *The Structure of Scientific Revolution*.

But even this situation is undergoing change, one that threatens the very future of a normative philosophy of science in the classical mold. On the one hand, the past ten years have been marked by some very important philosophical work in the foundations of virtually all the special sciences. On the other hand, there has been relatively little inspiration in the classical philosophy of science, with which the field is still popularly associated. We philosophers testify to this state of affairs whenever we try to outdo each other in the modesty of our theories of scientific rationality or apologize profusely for having to go through the motions of the scientific realism debates yet again. Not that rationality and realism are unimportant. On the contrary, arguing about these two issues routinely forces us to confront some fundamental problems about the nature of knowledge. Still, I would be the first to admit that arguing about rationality or realism for purposes of winning the debate has run out of gas. As might be expected under the circumstances, more and more philosophers are coming to believe that the future of the philosophy of science lies either in some other branch of science studies (especially history and sociology) or in the conceptual foundations of the special sciences. In fact, this future is already unfolding at our professional conferences. In the meeting rooms, the Old Guard are renouncing the errors of their classical ways, while in the corridors, the Young Turks are wondering, "What was all this fuss about Kuhn and Popper, anyway?"

I have no intention of denying any of these facts. However, I see their significance quite differently from many of my colleagues. To help the reader understand my perspective on the current state of play, I have provided a "map of the field" following this introduction. It is organized around the major schools within the two main claimants to the "science of science" or "metascience," namely, the philosophy and sociology of science. Some see this increased specialization, particularly within philosophy of science, as signs of maturity. I, on the other hand, see a withering away of the field, now that successive generations of philosophers have immersed themselves in the local knowledge of the special sciences, perhaps never to return to the more general debates that marked the philosophy of science in its heyday. I am not simply complaining about the perennial ingratitude of youth, but about the failure of recent philosophers to communicate to the larger academic community the concerns that they have developed in concert with local knowledge producers. In a sense, then, this book is a journey through these locales that aims to transfer their insights to an arena where they may have greater currency. After all, debates about causation that have by now taken a scholastic turn in, say, the philosophy of biology may substantially challenge, if not revise, our more general notions of science, when transferred to a more public arena.

From my standpoint, the biggest obstacle in the way of this free trade across disciplinary boundaries is exactly that part of the philosophy of science with which philosophers are most likely to infect the special sciences, namely, that illusory object of philosophical study, the *internal history of science*. The plausibility of the internal history of science has been supported by a certain selective history of the philosophy of science, one whose nineteenth and twentieth century focus is largely on philosophers writing in the English language (and their foreign emulators), rather than on those writing in French or German. By comparison with France and Germany, where disputes over disciplinary boundaries and science–society relations have been the norm, Britain remained, until quite recently, a nation of institutionally unaffiliated and isolated scientific geniuses (cf. Ben-David 1984, chs. 5–7). This difference crucially affected the terms in which philosophical disputes about the pursuit of knowledge were couched in these countries. It would be naive to think that the rather exclusive focus on the British side of things has not colored what we as philosophers of science take to be relevant to a philosophical understanding of science.

Indeed, I suspect that a symptom of our failure to come to grips with the Continental European philosophical tradition is the unsatisfactory treatments of facts and values, "naturalism" and "normativism," in scientific methodology. As I argue in Chapters Three and Four, these are very important matters about which we should have something interesting to say. Here the history of the social sciences, not the natural sciences, is a generally more helpful guide. In particular, I have recourse to the German Neo-Kantian debates over the study of humanity as part of nature (*naturwissenschaftlich*) or as a thing unto itself (*geisteswissenschaftlich*), as well as the *Methodenstreiten* over the scientific status of the social sciences and the relation between social science and social policy. In the end, my point is that the study of science should be conducted so as to be subsumable under a unified social science, which in its search for regularities and causal mechanisms will provide the basis for science policy. My general advice is that philosophers of science should take more of a third-person perspective toward the study of science, and think of the normative dimension on the model of political action rather than of aesthetic evaluation. I say all this, however, while dealing primarily with texts in analytic philosophy and sociology of science, including those focused on the most recent developments in cognitive science, evolutionary epistemology, and especially the experimental psychology of science. The influences from Marx, Foucault, Derrida, and Habermas are kept firmly in the background here. Readers who like to see their continental philosophy more up-front are advised to consult my first book. In contrast, the people who have provoked me the most to write this book are the ones I criticize the most, especially Larry Laudan who has dealt with these issues more openly than anyone else recently. Just how much I have been provoked can be gleaned from the table of contents, which reads as a spirited outline of my central argument.

My own philosophical stance is called *social epistemology*. In terms of the recent philosophy of science debates, my position is as follows. I am a scientific realist with regard to the discourse of the social sciences. By that I mean that the best explanation for the history of all of our knowledge enterprises is provided by the best social scientific theories. However, I am an antirealist about the discourse of the natural sciences, to the extent that I accept the validity of social constructivist accounts of natural scientific practices. In particular, I accept the constructivist conclusion that the practices of natural scientists, when judged on the scientists' own terms, are only locally explicable. However, I do not accept the further constructivist claim that there is no way of getting a more comprehensive "scientific" understanding of what the natural scientists are doing. There is such a way, but it involves importing the categorical framework of the social sciences and recasting what the natural scientists are doing as instances of more general forms of social behavior. One of the consequences of this move is to show that the discourse of the natural sciences misleads its speakers in systematic but generally productive ways. Thus, I have a rather robust sense of Hegel's Cunning of Reason acting in the history of science. In fact, it is the sort of rationality that is probably the most interesting to pursue, which explains why I continually stress in this book the dialectical interplay between the first-person perspective of the scientist and the third-person perspective of the historian or policy maker. The problem of rationality, as I see it, is the problem of resolving these two perspectives. This is where such humanistic intrigues as ideology critique and deconstruction, with their close attention to the history of language, can be most helpful. (This is explored more in Fuller 1988b.) Finally, since I believe that this entire analysis applies to the social sciences and humanities as well, I must be a fallibilist, which is to say, I believe that the social epistemologist should accept the likelihood that her own analysis is also ultimately false, but I hope, false in ways that prove instructive to her successors.

I come to social epistemology by way of a few homely observations. Knowledge exists only through its embodiment in linguistic and other social practices. These practices, in turn, exist only by being reproduced from context to context, which occurs only by the continual adaptation of knowledge to social circumstances. However, there are few systemic checks for the mutual coherence of the various local adaptations. If there are no objections, then everything is presumed to be fine. Given these basic truths about the nature of knowledge transmission, it follows that it is highly unlikely that anything as purportedly uniform as a mind-set, a worldview, or even a proposition could persist through repeated transmissions in time and space. Consequently, unless this point has been explicitly taken into account, if a philosopher or historian claims to have isolated such a uniformity, we should infer that it is an artifact of her analysis, a sign that she has failed to resolve her abstractions at the level on which history actually occurs.

Although philosophers nowadays nominally admit the contextual character of knowledge, we are still very much in the habit of thinking that science is distinguished from other forms of knowledge by its ability to preserve propositions, or some other notion of *content*, across contexts. In fact, most philosophies of science are devoted to explaining how this purported preservation of content is possible: Is it that scientific reasoning follows a deductive pattern of inference, or that the ultimate court of appeal in science is repeatable observation, or something else entirely? And although the explanations vary widely, they all agree on the phenomenon that is to be explained. This testifies to the sway of the Metaphysics of Inertia: Continuity is the way things are naturally, unless they are interrupted. (By contrast, I hold to the Metaphysics of Entropy: When continuity is not enforced, discontinuity reigns.) This should come as no surprise, given what is at stake here. If content were not normally preserved, then it would make no sense to speak of "adding to" or "subtracting from" the storehouse of knowledge; an implication of either idiom is that knowledge remains contained in its linguistic packaging unless explicitly changed.

To see the disconcerting consequences that this view has for the philosophy of science, consider how one might now go about explaining the connection between a theory *being true* and the theory *appearing true* to several scientific communities. Since the contexts of appearance, so to speak, are generally in rather diverse social circumstances, philosophers have been inclined to suppose that the theory must have some content, or truth, that remains invariant across these circumstances. Moreover, as we have just seen, philosophers normally assume that continuity of this sort is necessary for the transmission and growth of knowledge. Yet, all this talk of "invariance" and "continuity" typically confuses the uncontroversial claim that the truth itself does not change with the dubious claim that the truth is transmitted intact by reliable linguistic means. Indeed, if it can be shown that we have less than reliable linguistic means for transmitting truths, then whatever invariance we seem to find in scientific theories that have been accepted across many sociohistorical contexts (e.g., Aristotelian cosmology, Newtonian mechanics, Darwinian biology) cannot be due to the invariant nature of the truths transmitted, *even if the theories are indeed true*. Rather, it must be due to institutionalized cognitive mechanisms that inhibit scientists from perceiving the real differences in interpretation that are a natural consequence of the theories' being unreliably transmitted. These mechanisms are probably of two basic sorts. One sort makes the scientist a sufficiently gross receiver of information that she ignores most of the real variance in the system. (Traditionally, this feature has been given a more positive gloss, in terms of the scientist's perception being focused by theoretical commitments.) The other relevant sort of mechanism enables the scientist to respond to the information she receives only within a narrow range of communicative options (e.g. disciplinary jargons and writing formats). Together these mechanisms define the sense in which a scientist truly undertakes a "discipline" in the course of her work.

Thus, I differ from those sociologists of science who would primarily appeal to differences in local interests to explain variation in knowledge production practices. (Not that these interests do not exist, but rather, that they explain little here.) By contrast, I hold that even if all scientists agreed on a set of interests to push as a social class, as long as the scientists remained as numerous and dispersed as before, most of the variation in their practices would still remain. This is because the source of the variation is something that is virtually impossible to avoid, given the current spatiotemporal dimensions of knowledge production. Thus, to engage successfully in a scientific discipline is simply to learn to minimize the impact of this brute fact on one's own activity. To find out how one does this and how it all adds up to an enterprise that seems to produce knowledge, please read on.

My Map of the Field

1. Overall Trend: From Historicism to Naturalism

Among the most influential academic books in this century is a philosophical treatise about the history of science written by a physicist who spent most of his time teaching general education courses. I am describing, of course, Thomas Kuhn's *The Structure of Scientific Revolutions*. The most palpable consequence of the enormous debate that surrounded this book in its first fifteen years of publication (1962–77) was the large number of courses and books on "methodology" that sprang up throughout the arts and sciences, which featured discussions of Carl Hempel, Russell Hanson, Stephen Toulmin, Mary Hesse, Karl Popper, Imre Lakatos, and Paul Feyerabend—philosophers who stood in an interesting dialectical relation to Kuhn. (Lakatos & Musgrave 1970, Suppe 1977, and Hacking 1981a are the best collections. The first half of Hacking 1983 gives an astute retrospective appraisal.) Most of these philosophers were originally trained in one of the special sciences, usually physics, and many fell under the spell of that arch-antiphilosopher, Ludwig Wittgenstein, at some point in their careers. That alone should suggest that these philosophers did not view philosophy as primarily worth pursuing for its own sake, but only insofar as it provided guidance for the conduct of inquiry in general.

In vulgar academic parlance, the more one disagrees with Kuhn, the more of a "positivist" one is. In this context, a positivist is someone who wants to derive universally applicable formal principles for the conduct of inquiry. Kuhn is then taken to have shown that the positivist project is completely misguided, given that the history of science shows that no such principles are to be found. Despite the grain of truth contained in this caricature, it ironically misplaces the logical positivists' motivations. In fact, the positivists were themselves ex-scientists who realized that the justification of knowledge claims was too important a problem to leave in the hands of professional philosophers. In that respect, Kuhn succeeded where the positivists had failed, which, in turn, vindicates the positivists' original sponsorship of Kuhn's book as part of the International Encyclopedia of Unified Science. But just because Kuhn succeeded in galvanizing the larger academic community, it doesn't follow that his readers took to his example. Indeed, it is chic (and maybe even justified) for professional historians to dismiss Kuhn as methodologically naive. And certainly, no *historian* after Kuhn has tried her hand at divining philosophical lessons from the history of science—though some philosophers have: Lakatos, Laudan, and Shapere figure prominently in this book. Still, Kuhn has had his uses, especially among theorists in the social sciences, who have issued programs for "normalizing" their fields at an alarming rate (Gutting 1979). By the mid-1970s few disciplinary practitioners felt legitimate unless they had a "paradigm" they could point to.

Some philosophers followed Frederick Suppe's (1977) lead in believing that the massive critical attention focused on Kuhn's book signaled the dawn of a new interdisciplinary field, "History and Philosophy of Science" (or HPS), a sort of humanist's revenge on logical positivism. Accordingly, close textual analysis of significant texts in the history of science would replace the grandiose claims to universal methodological norms that marked the earlier positivist regime. Suppe's negative claim certainly turned out to be right, in that the only widely discussed book to extend the positivist program since 1977 is Glymour (1980). And, like many good positivists, Glymour has been driven in more recent years to conceive of his project as a form of artificial intelligence, specifically an epistemology for that perfect formal reasoner, the computer android. Indeed, it would be fair to say that the positivist impulse is alive and well in cognitive science.

Where Suppe was wrong, however, was in his positive claim that HPS would dominate the research agenda of philosophy of science. Suppe's exemplar was the sophisticated historicism of Dudley Shapere (e.g., 1984), whose attempt to articulate an "internal history of science" animates the critical side of the book before you. Nersessian (1987) presents a variety of recent philosophers who have drawn inspiration from Shapere. However, only Nersessian (1984) herself has tried to retain Shapere's grand vision of historically derived, yet transhistorically applicable, methodological principles. But, in her hands, his vision takes the form of a normative cognitive psychology of science that very selectively draws from the conceptual and methodological resources that psychology currently has to offer (cf. Fuller 1991a). This point becomes especially clear when Nersessian's work is seen in light of what is still the most cosmopolitan attempt to integrate psychology into HPS (De Mey 1982), as well as the first two volumes of case studies on scientific discovery (Nickles 1980a). In general, however, it has not been easy to find a principled way of allowing empirical research in the social sciences to inform normative theories of scientific rationality. This is perhaps the biggest problem—how to license the move from "is" to "ought"—that plagues the various philosophers who travel under the rubric of "naturalism" (cf. Fuller 1992b).

Sad but true, then, HPS has gradually lost its momentum. The younger generation of historians of science are the bellwether figures here (e.g., Shapin & Schaffer 1985, Porter 1986, Galison 1987, Dear 1988, Proctor 1991). To a person, they define themselves explicitly, if sometimes oppositionally, to the theoretical agendas of the sociology— not the philosophy—of science. (A common publishing outlet is the journal, *Science in Context.*) This point is even reflected in the formerly staid *Studies in History and Philosophy of Science*, whose latest editor, Nicholas Jardine, has himself written what is probably the most sociologically progressive book to come out of the HPS camp (Jardine 1991). On the surface, it may seem strange that historians would turn to sociologists instead of philosophers for theoretical sustenance. But this just reveals the extent to which the debates surrounding Kuhn only *temporarily* halted the onrush of specialization within philosophy. Here it is worth recalling a point that always seems to elude philosophers, especially epistemologists—whom I will keep separate from philosophers of science for purposes of this discussion. The abstractness or generality with which one talks about knowledge does not ensure that what one says will be of relevance to the larger academic community. Thus, people who call themselves "philosophers of science" have had wider impact than people who call themselves "epistemologists," even though epistemologists do not officially limit their study to disciplined knowledge, or science.

Whereas epistemologists want to know when an individual's belief ought to count as knowledge, philosophers of science want to know when a theory ought to be accepted by a community of researchers. Interestingly, epistemologists tend to treat commonsense and science as making equally valid (or invalid, if one is a skeptic) claims to knowledge, while philosophers of science presume that science is superior to commonsense and maybe even good enough to defeat the skeptic. As a result, epistemology often appears to be a static enterprise, one devoted to elaborating timeless criteria for knowledge, in contrast to the more dynamically oriented philosophy of science, which is full of arguments about one theory replacing another. It is worth noting that this dynamism applies just as much to the positivist preoccupation with later theories logically subsuming earlier ones (i.e., "reductionism," e.g., Nagel 1960) as to the post-Kuhnian focus on later paradigms radically displacing earlier ones.

With a few notable exceptions (e.g., Hacking 1983 among philosophers of science, and Rorty 1979 among epistemologists), epistemologists and philosophers of science are united in a common normative interest—in what people ought to believe—which has classically been treated as an a priori question, one that must ultimately be defended on the basis of conceptual arguments alone, no matter what the actual empirical record of science might suggest. The move to "naturalize" epistemology and the philosophy of science involves injecting facts about our actual epistemic situation into these normative considerations. Yet, in practice, this move tends not to be as radical as it seems (e.g., Kornblith 1985).

As a matter of fact, naturalism had reached its peak early in the twentieth century (especially among such pragmatists as Peirce, Dewey, and Mead), when philosophers had no qualms about evaluating science on the basis of its liberating social consequences. But once science started to function as the motor of military and industrial technology, people began holding science accountable for much of the danger and destruction in the modern world. This has led pro-science philosophers—as philosophers of science invariably are—to draw a sharp line between a scientific theory and its technological applications. Consequently, the more modest naturalist of today will either use an a priori conception of knowledge to determine which historical or psychological facts are relevant to epistemology (e.g., Goldman 1986, 1991) or she will use such facts to determine the range of possible norms that warrant further conceptual investigation (e.g., Laudan 1987). In both cases, naturalism only reduces the number of possibilities from which the philosopher derives norms for knowledge, but it does not dictate any specific solutions. However, this book's heroes are the philosophers who defend a more radical form of naturalism, one which makes epistemology and the philosophy of science continuous with science itself.

In commenting on more radical naturalists, I shall merely mention the work of the distinguished but underrated Australian philosopher, Clifford Hooker (1987), who gives evolutionary epistemology a Leftist political spin, courtesy of cybernetics. Part of Hooker's relative neglect is due to Anglo-American philosophers having gotten so used to keeping issues of politics and science apart that it rarely occurs to any of us that politics may be part of the essential nature of science (and vice versa, as Ezrahi 1990 maintains for contemporary Western democracies). At least, the true naturalist is under an obligation to keep this possibility empirically open, not conceptually closed. Moreover, given the politics-as-nature thesis that I have just suggested, our naturalized inquiries into the nature of science may leave us with a somewhat changed attitude toward both our object of study and the methods we use to study it. This is the "reflexive" challenge that is characteristic of the most challenging recent thinking in the sociology of knowledge, as discussed in the next section.

However, for most philosophers of science, it is radical enough to claim that philosophical problems are best solved—or at least developed in a promising direction—by the tools of the social sciences. This book explicates one version of this thesis. Here it is worth comparing my view with that of the other philosopher who takes this tenet of naturalism as seriously as I do, Ronald Giere, especially in *Explaining Science* (Giere 1988). Of relatively little import is the difference between, so to speak, my "realistic constructivism" and Giere's "constructive realism" as an overall philosophy of science. More important is an ontological difference: Giere adheres to the ontological primacy of individuals and I to that of collectives. Thus, Giere tends to treat the social psychology of science as the aggregation of the individual (specifically cognitive) psychologies of scientists, while I stress the emergent, though still largely cognitive, character of the social. Thus, I hold that "knowledge" in the sense that has traditionally interested philosophers (i.e., theories, research programs, etc.) is not only the product of social interaction but is also itself something of which an individual could possess only parts but not the whole. A sense of the difference between the research agendas of Giere's supporters and my own may be gotten from the impressive fifteenth volume of Minnesota Studies in the Philosophy of Science (Giere 1992).

The best evidence I have of naturalism's relatively undertheorized condition is that Giere and I make two sorts of assumptions that could probably benefit from some argumentation. First, while we are both strongly in favor of empirically testing the normative theories that philosophers have proposed to govern science, we seem to differ with regard to the sort of test that is appropriate. Giere seems to lean toward history, whereas in this book I lean toward experiments. However, this is a relative difference, as both of us rely heavily on a broad range of results from the history of science, experimental psychology and cognitive science. (An exclusively historical but more methodical approach to such tests has been pioneered by Larry Laudan, especially Laudan et al. 1986 and Donovan et al. 1988.) Secondly, a naturalist must be prepared to face the prospect that the empirical findings underwriting her philosophy of science may not be compatible with one another, especially if they are drawn from different disciplines, such as psychology and sociology. How does one resolve this issue in a way that transcends the self-serving? As it stands, whereas I tend to accept sociological data at face value and reinterpret psychological data accordingly, Giere tends to do the opposite. No surprise there.

Readers interested in pursuing the naturalistic project in the philosophy of science would be wise to look at the following works. Tweney et al. (1981) and Gholson et al. (1989) are collections that range over, respectively, historical and contemporary concerns in the psychology of science. Tweney's volume is divided according to the topics that have traditionally concerned philosophers, which means that the lion's share of the discussion is devoted to the contexts of "discovery" and "justification." Gholson's volume compensates for this with extensive coverage of social psychology. Since "social psychology" is a field that falls under both sociology and psychology, some interesting contrasts of perspective can be expected. Sociologists and psychologists of science first locked horns in the 1989 Sociology of the Sciences Yearbook (Fuller et al. 1989). Shadish and Fuller (1992) have since constructed a wide-ranging research program in the "social psychology of science." Those interested in the relationship between experimental methodology and human cognitive limitations, and its bearing on the construction of normative theories of science would benefit from the philosophically literate treatments provided by David Faust (1984) and Arie Kruglanski (1989). Since Faust figures prominently in the main body of this text, let me say something here about Kruglanski.

According to Kruglanski's program of "lay epistemics," people are endowed with logical thought processes, but these are rarely realized in practice because the beliefs on which those processes operate are either false or, if true, the beliefs are not available to the agent on a reliable basis. The latter especially wreaks havoc on people's ability to learn from mistakes because we very often hold the right beliefs, but due to a limited access to those beliefs, we are unable to apply them to draw the right conclusions. As a result, *acknowledging error* (in the past) and *avoiding error* (in the future) turn out to be two disturbingly different sorts of activities. Under these circumstances, experimental design is important in providing the compensatory cognitive strategies that are needed for any genuine growth of knowledge to occur. These strategies are less formulae for discovering the truth than heuristics for detecting the more pernicious errors to which the *inquirer* is prone—especially the tendency to participate in the illusions of the subjects she studies! Here Kruglanski has distinguished a "realist" paradigm of experimental accuracy, based on holding constant the experimenter's original objective (often by a non-human measuring device), and the "phenomenalist" paradigm of accuracy that subjects themselves would use in everyday life. This distinction bears critically on the use that sociologists of science (see below) make of the pragmatist motto so frequently invoked by naturalists—that the validity of a belief rests on the belief's consequences for action (cf. Fuller 1992b). Whereas most philosophical naturalists prefer the realist paradigm associated with Peirce, sociological naturalists, as we shall see, tend toward the phenomenalist paradigm associated with William James.

2. The Great Pretender: The Sociology of Scientific Knowledge

The year 1977 marked the publication of the book that was supposed to have launched the Kuhnian Revolution into its second phase, Larry Laudan's *Progress and Its Problems*. Laudan initially earned his reputation as an astute historian of the philosophy of science, and Laudan (1982) is still the best set of essays on the eighteenth and nineteenth century origins of current debates. However, in *Progress*, Laudan tried to reconcile the most extreme tendencies of Kuhn and his positivistic foes. On the one hand, Laudan was willing to bite the bullet and grant Kuhn that alternative paradigms (which Laudan calls "research traditions") cannot be compared. But, this still leaves open the possibility of traditions being compared in terms of how well they solve the problems they set for themselves. Laudan's standard is that traditions aim to solve a certain set of self-specified empirical problems without generating too many conceptual ones in the process. Yet, on the other hand, Laudan wanted to revive Reichenbach's (1938) positivistic conception of rationality, which declared sociologists and other non-philosophers relevant to the enterprise of explaining science only when scientists violated philosophical canons of rationality.

This bald reassertion of philosophy's priority was immediately challenged by a new breed of mostly British scholars who rallied around "The Strong Programme in the Sociology of Knowledge," an expression coined by David Bloor (1976) in *Knowledge and Social Imagery* and further elaborated in Bloor (1983). Both works contain some bold attempts at explaining that most intellectual of intellectual pursuits—mathematical discovery—in sociological terms. Counterposing a "Left Kuhnian" interpretation to Laudan's "Right Kuhnian" one, Strong Programmers argue that since "science" primarily picks out a set of social practices, it can be studied scientifically with the same categories and tools one would use to study any other social practice. No pride of place is given to explanations that impute a form of rationality that only scientists have, and no one else. The Strong Programme's basic methodological move is to portray knowledge claims as symbols standing for, and manipulated by, competing groups in a highly structured agonistic field. The most sustained historical treatment in this vein is Shapin & Schaffer (1985), which analyzes the social ascendency of experimental knowledge in seventeenth century Britain. An equally sustained treatment of recent Big Science is Pickering (1984). Bloor and his colleagues at the Edinburgh University Science Studies Unit founded a journal in 1970, *Social Studies of Science*, which remains the principal organ for this kind of research.

It is worth noting that the "social" in, say, "social studies of science" should not be taken as a disciplinary commitment to sociology. After all, neither Bloor nor his colleagues were trained in the field, and, indeed, anthropological methods have come to dominate the field. Rather, "social" refers to a more general metaphysical commitment to "sociologism," much as a philosopher might want to call herself a "materialist" without feeling the need to endorse any particular theories in the physical sciences or its current division of disciplinary labor. The relevant precedent here has been set by Marxists, who have typically been materialists yet, at the same time, quite suspicious of microphysical inquiry as diverting scientific attention from the level at which the laws of political economy operate (cf. Aronowitz 1988).

What makes the Strong Programme "strong" is its commitment to demonstrating that not only the rationality of scientists, but the very content of scientific knowledge could be treated by social scientific methods. This was stronger than most sociologists were willing to allow themselves. For, in consigning the sociology of knowledge to the "arational" side of science, Laudan had simply followed the advice of that field's founder, Karl Mannheim (1936). Indeed, although nowadays it is natural to think of the sociology of science as a branch of the sociology of knowledge, this is a development that is due largely to Robert Merton (1977). Mannheim argued that since the sociology of knowledge relativizes the justification of knowledge claims to the cultures that sustain them, science, because of its universal truth claims, could not itself be an object of sociological study. Otherwise, sociology of knowledge, as a scientific discipline, would be reduced to mere ideology. Against this, Merton observed that even the search for truth requires a specific kind of (democratic) social order, which he captured in terms of norms that reflected what philosophers have traditionally taken to be the virtues of the scientific method. Thus, Merton included rationality but not content within his purview.

Despite their differences, Merton and Bloor agree that, in some sense, the science and society of a given period are reflections of one another. Whereas Merton stressed the way in which society is organized to facilitate scientific inquiry, Bloor highlighted the extent to which scientific disputes are social struggles in symbolic disguise.

However, once the sociologists started to conduct ethnographies of science as it was actually being done in laboratories, they discovered a massive disparity between the words and deeds of scientists. The words, as they appeared in journal articles, largely conformed to philosophical canons of rational methodology, but they also represented a highly idealized—if not downright misleading—picture of what took place in the lab. The lab work itself turned out to be quite chaotic and open-ended, even at the level of personal interests and group understanding of the ends of their research. The two major monographs in this vein are Latour & Woolgar (1979) and Knorr-Cetina (1980). Collins (1985) is a series of studies that challenge particular method–practice disparities, while Gilbert and Mulkay (1984) offers a guide to the analytical tools used in ethnographic studies of science. Together these works constitute the foundations of the emerging field of history, philosophy, and sociology of science known as Science & Technology Studies (STS). Newcomers to this field can quickly get up to speed by reading Woolgar (1988b). State of the art research and debate is presented in Pickering (1992).

A good way of understanding the evolution of STS is through the career of Michael Mulkay (esp. 1990), typically regarded as the founder of the more radical sociology represented in STS, *social constructivism*, whose methodological imperative is to regard science as being made up as the scientists go along. Mulkay only gradually came to realize that scientific practice does not live up to its own hype. Early in his career, Mulkay noticed that Merton substantiated his claims about the normative structure of science largely on the basis of the remarkably uniform testimony of scientists and philosophers. But uniformity of word does not imply consistency of deed. Mulkay was struck by Merton's failure to operationalize such avowed norms as "universalism," "communalism," "disinterestedness," and "organized skepticism" so as to examine the extent to which they had a purchase on the day-to-day activities of scientists. For, upon closer scrutiny, Mulkay discovered that a host of mutually incompatible practices have been justified by appealing to the same set of norms. In other words, what sociologists had identified as the normative structure of science really turned out to be a set of rhetorical resources that scientists routinely used to justify whatever they happened to be doing at the moment. The norms themselves had little predictive or explanatory power.

From these skeptical origins began the constructivist turn in the sociology of science. Mulkay's work has breathed new life into such moribund philosophical concerns as theory choice and consensus formation in science by showing that whatever certainty we seem to have about whose side won in a particular scientific debate is more the result of the winner's ability to suppress alternative accounts of the event than of some knock-down empirical demonstration. A particularly salient application of this point appears in Mulkay's (1990, ch. 8) important essay, "Knowledge and Utility," in which he systematically debunks claims that scientific breakthroughs either cause or explain the major technological successes of the last 150 years. The case against causation is easily made because technologies generally predate the scientific knowledge that is supposedly required for their invention. Moreover, the point applies not only to commercial technologies but also to medical ones, such as treatments for disease, which often turn out to contribute little to life expectancy beyond what had been achieved by improvements in public hygiene. The case against science explaining technology is a little trickier, but the key here is to notice the ease with which we credit science with, say, a successful space mission but fault "human error" when the spacecraft crashes. Science seems to have such a good track record in explaining technology because scientists enjoy a monopoly over accounts of the successes and can exercise discretion in laying blame for the failures.

Mulkay's work easily leaves the impression that if scientists are expert in anything, it is in the fine art of "spin doctoring." But while an insight of this sort invites a certain cynicism toward the scientific enterprise, it can also be used strategically to demystify appeals to scientific authority that are made in order to close off public debate. In this spirit, Mulkay has recently turned his attention to the efforts of "health economists" to make more precise (i.e., more quantitative) ordinary intuitions about the quality of life and, in the process, usurp from patients the right to speak with authority about their own state of health. By employing "new literary forms," Mulkay tries to make space in his own text for the patient to address the inquiring sociologist. (Woolgar 1988a is a convenient site for examining the aims and products of this recent development.) The result typifies the strengths and weaknesses of constructivism, both in its ability to tease out structures of domination from discourse *and* in its inability to offer an alternative politics of science. However, as Mulkay himself suggests on several occasions, perhaps the point here is that policymaking is not exclusively in the hands of the sociologist, but requires the participation of the scientists under investigation and the potentially affected third parties.

Mulkay's recent forays into health economics is part of the general constructivist tendency toward constructing a *sociology of representation*. Much of the hottest work in this field (e.g., Lynch & Woolgar

1990) is devoted to identifying the interpretive conventions that enable scientists to find certain models and visual displays so "realistic" that they no longer feel the need to deal with the realities that such pictures, graphs, and simulations represent. The important point to realize here is that, whether they are studying narrative or graphic devices, the constructivists understand "representation" in a very broad way that even includes the political senses of the term.

Normally, we think of "representation" in the political and the "semiotic" (i.e., linguistic and pictorial) senses as mere homonyms, an instance of the same word being used to mean two different things. However, one need only turn to Aristotle and Hobbes to see that this distinction in meanings is a relatively recent innovation, one which the sociologists are keen to reverse, at least for their critical-analytic purposes. The unifying idea here is that when one thing or person, X, represents another, Y, three things happen at once:

(1) X speaks in the name of Y,
(2) Y no longer speaks for itself, and
(3) there is no longer any need to refer to Y.

Just as the health economists claim to "represent" the interests of patients, so too do pictures, graphs, and simulations purport to "represent" a reality that scientists can then officially do without. To show how the process of representation works is to articulate the ways in which even our ordinary understanding of the world is mediated by scientific structures. And so, while science may ultimately not be responsible for splashy technological effects, it nevertheless exerts a subtler and more pervasive influence in shaping our sense of, say, what a "good" drawing of an animal looks like.

While constructivists loudly advertise the radicalism of their claims, it may not be clear what all the fuss is about. For some clarity, think of constructivist accounts of representation as standing to more orthodox realist accounts much as Lavoisier's oxygen stood to Priestley's dephlogisticated air. Corresponding to the pseudo-process of dephlogistication is what we normally think of as "decontextuali-zation," namely, the ability of an especially realistic representation to abstract essential qualities from a variety of contexts so that an object is made plain for all to see. But the sort of representation that counts as realistic will depend on what the object is taken to be. So, while an individual human may be realistically represented by a photograph, an individual animal may be treated to a detailed sketch, whereas stylized drawings may lay claim to realistically representing either parts of an animal or an entire ecology.

In any case, constructivists replace the occult process of decontextualization with a sequence of operations that serve to standardize the object for the relevant scientific audiences. Rarely is the sequence exactly the same, as indicated by the disparate accounts that scientists give of what exactly makes a particular representation realistic. Yet, certain texts tend to be juxtaposed in journal articles as captions and commentaries for the representation, and certain routines tend to be performed by scientists in the presence of the representation. But most importantly, scientists in a particular research setting come to be convinced that scientists in other settings treat the representation exactly as they do. Although scientists generally know better than to believe in the existence of such uniformity, nevertheless the verbal and visual representations assembled in the name of science manage to sustain an image of a universally accessible reality.

The distance that the social construction of scientific imagery takes us from rationalist philosophy of science reveals the irony in the acrimonious debate that ensued between defenders of Laudan and Bloor (compiled in Brown 1984), for they disagree over much less than they think. In the first place, both fancy themselves in search of a "purified" understanding of science. Echoing the later Wittgenstein, Bloor holds that science can be explained as a form of life without importing conceptions of truth, rationality, and reality that require special philosophical grounding. In fact, historically speaking, Strong Programmers argue that the appeal to philosophical concepts has largely had political, not scientific, import—to consolidate allies and to exclude rivals. Needless to say, this interest in purging ideology from science also animates Laudan's antipathy to the sociologists, as well as his aversion to the metaphysical pretensions of scientific realists! In addition, being a self-styled "pragmatist" (Laudan 1990), Laudan holds a consensus theory of validation and an instrumentalist theory of rationality, which puts him intellectually much closer to the sociologists than to most of his fellow philosophers—a point brought out in reviews of Laudan (1984).

It may be, then, that more heat than light is shed by the very distinction between "philosophical" and "sociological" approaches to science. It really *is* a political battle over who has the right to speak for, or "represent," science, with all the benefits that are understood to accrue to its representative. This is a recurrent theme in the work of the philosopher-turned-sociologist, Bruno Latour. As Latour (1987a) puts it in *Science in Action* (perhaps the most popular book in STS today, known for its jokes almost as much as for its arguments), Laudan and Bloor are simply doing what scientists are always doing—only the scientists do it for larger stakes. (A student of Michel Serres, Latour

[1989] articulates the Neo-Lucretian metaphysics of the "agonistic field" that underwrites his conception of science.) In a more conventionally philosophical vein, Paul Roth (1987) has hoisted Bloor and company by their own philosophical petard in the first full-length appraisal of the Strong Programme. Interestingly, Roth complains that the sociologists retain too much of the old positivism for their own good. All this goes to show that the fighting is fiercest between like-minded thinkers in rival disciplines.

3. The Old Chestnuts: Rationalism and Realism

As the dust starts to settle on the Laudan–Bloor debate, it is becoming clear that the sociologists have made some subtle inroads into philosophical thinking about science. True, there are still few philosophers who endorse anything like the Strong Programme. But now, there are also few philosophers who try to defend global claims about science being uniquely "rational" or "realist" in its methods. Indeed, there is growing philosophical suspicion that rationality and realism—traditionally regarded as the core issues in the philosophy of science—may turn out to be little more than licenses for reconstructing the history of science in order to pass judgment on today's scientists. This suspicion has been given eloquent expression, for the natural sciences, in Hacking (1983), and, for the social sciences, in Ackermann (1985).

In brief, for someone to be deemed "rational" or for something to be deemed "real" is to say that they ought to be left alone. However, this normative injunction is typically disguised as a claim to ontological status: the rational person is "autonomous," and the real thing is "independent." An indication of the extent to which philosophers of science have retreated from their earlier global aspirations is that when they nowadays speak of the "rationality debates" and the "realism debates," they usually mean debates about the foundations of specialties in the social and physical sciences. Roughly speaking, in the social sciences, rationality is what under-writes the reality of people as proper objects for those sciences, whereas, in the physical sciences, rationality is what underwrites the trajectory of research in those sciences as getting ever closer to the nature of reality. In both cases, rationalists and realists have their work cut out for themselves.

Social scientific realists take the proliferation of social science disciplines and methods as a "problem" that requires systematic treatment. Like their unity-mongering forebears, the logical positivists, realists have characteristically identified a genuine issue that philosophers are well suited to address because of their relative detachment from the disciplines concerned. Unfortunately, their "cure" threatens to eliminate ontologies that presume that to each social science department there corresponds a unique domain of objects (i.e., psychological entities, sociological entities, economic entities, political entities, etc.). From my own standpoint as social epistemologist, this is an unnecessarily intimidating way of addressing what is more likely to be the result of discourse communities that have become self-contained because they have had no need to confront each other's knowledge claims. The point, then, would be to develop a rhetoric that cuts through these essentially linguistic, not ontological, barriers. To reiterate the main thesis of Social Epistemology (Fuller 1988b), major conceptual difference is due to communication breakdown, which, if it goes on long enough, will simulate the sort of "incommensurability" that originally led Kuhn to say that scientists in two different research traditions lived in "different worlds."

Some realists are worse offenders than others. After surveying the central debates in each of the social sciences, Alexander Rosenberg (1989) draws the perverse conclusion that the collective failure of these disciplines is a sign that homo sapiens does not constitute a natural kind, and hence is "by nature" precluded from becoming the subject of a real science. While Rosenberg embraces sociobiology as the way "beyond" the morass of the social sciences, Peter Manicas prefers a more charitable solution. Manicas (1986) argues that the social sciences got off on the wrong foot by buying into the positivistic misunderstanding that the natural sciences had of themselves. Thus, social scientists felt inadequate unless they could predict and control behavior. Manicas is clear that the problem is not the idea that the natural sciences are superior knowledge enterprises—the natural sciences justly earned their reputation with the explanatory unity of Newtonian mechanics, as it promised access to some real underlying mechanisms. Manicas then shows how this perspective can alleviate many of the conceptual problems normally seen as specific to the social sciences, which he thinks arise only because of hidden positivist assumptions.

The "kinder-and-gentler" realism espoused by Manicas belongs to the peculiar breed of non-positivistic realism that is grown at Oxford these days. Rom Harre (1970, 1986) and his student, the Marxist Roy Bhaskar (1979, 1980, 1987), are the leaders of this movement, which attempts to revive what I call in this book an "Aristotelian naturalism." Although they are officially very friendly to the social sciences, they also believe that social scientists typically do not appreciate that they are studying something whose essence ("rationality" again) is fundamentally different from that of natural things. To put their point as a paradox, what enables us to successfully intervene in the course of nature through experimentation is also what prevents us from ourselves being studied by that method (cf. Harre & Secord 1982). In particular, Harre (1979, 1984) believes that taking this point seriously will foster a science of *moral agents*, not simply of people-like animals. This move, regularly documented on the pages of *The Journal for the Theory of Social Behaviour*, is typical of continuing philosophical attempts to delimit the scope of the social sciences on a priori grounds (Hollis & Lukes 1982 surveys the array of strategies). By contrast, theoretically inclined social scientists tend to believe that key epistemological assumptions are tied, not to the nature of their subject matter, but to the particular method of empirical inquiry they happen to adopt (cf. Fiske & Shweder 1986).

Two international journals, the Canadian *Philosophy of the Social Sciences* and the Scandinavian *Inquiry*, are the leading forums for debate between philosophers and theoretically inclined social scientists. However, in recent years, economics and technology have developed into relatively autonomous philosophical specialties. There is substantial historical precedent for these two specialties evolving in conjunction with one another. After all, Karl Marx, Thorstein Veblen, and especially Joseph Schumpeter viewed technology as the source of increased productivity in the capitalist economy—albeit with often self-defeating long-term consequences. This tradition continues to be well represented in the work of the ingenious and prolific Norwegian political theorist, Jon Elster (1983), which argues that economic models of the spread of technological innovations can teach philosophers of science a lot about the processes that enable conceptual change to occur more generally.

However, most self-styled philosophers of technology resist any easy assimilation of the technical to the economic—or to the scientific, for that matter. Agassi (1985) is a good case in point. He stresses that only in relatively recent times have scientists been directly involved in the construction of technology, and that, in many instances, this involvement has served to dull the scientists' critical sensibilities, as well as those of the general public, who ostensibly benefit from the technology. Rapid technological progress can undermine the democratic spirit while improving the material character of our lives. Much recent sociology of technology in STS supports Agassi's viewpoint, in its attempt to open the "blackbox" that technology tends to become, once it has left the designer's hands and enters a diffuse distribution network (cf. Bijker et al. 1987). A more optimistic vision is projected by the phenomenologist Don Ihde (1987), who shows how technology has enabled the construction of a plurality of environments between which we can move in and out. Ihde (1991) brings this insight back home to the philosophy of science, as he focuses on the overwhelming role that apparatus plays, even in defining what the physicist nowadays counts as "physical reality." Finally, drawing on Heidegger and the hermeneutical tradition, Heelan (1983) has provided the most sophisticated development of this insight, whereby experimental observation becomes quite explicitly a means of "reading" nature.

In turning to physics, the most obvious thing to say is that when philosophers suppose that the success of "science" is best explained by its special access to reality, they really mean the most advanced science, physics. It is important to realize, though, that this fixation on physics is largely a twentieth century preoccupation, resulting from the field's remarkable resilience after the revolutions wrought by relativity and quantum theories. In the nineteenth century, physics interested philosophers because of the possibility that its methods could be transferred to disciplines that have yet to fully mature. Physics itself, on the other hand, was then seen has nearing the close of its inquiries. As the most "advanced" science, physics was treated as more senescent than progressive. The logical positivists changed all that, as they were among those most impressed by the reformation of physics in the first decades of this century. Moreover, these first impressions did not wear off on Popper, Kuhn, and their followers, however much they disagreed with the positivists on other matters. The work of Bas van Fraassen remains closest to this spirit, especially *The Scientific Image* (van Fraassen 1980), which must be credited with reviving interest in whether the success of physics is due to its having captured something about the nature of reality. The sustained negative answer given in that book has elicited a variety of realist responses (e.g., Leplin 1984a, Churchland & Hooker 1985), all of which goes to show that even if physicists were to converge on the nature of reality, realists would still need to converge on the conception of reality that explains what the physicists had done!

Since the mid-1980s, debate has increasingly focused on the conceptual foundations of particular branches of physics. Here an effort is typically made to recover historical considerations that originally led both philosophers and physicists to think that there was something special about the sort of knowledge produced by physics. Interestingly, these works emulate van Fraassen's strategy of seeking legitimization from such physicist-philosophers as Pierre Duhem and Ernst Mach—all the while downplaying Duhem's and Mach's own normatively inspired readings of the history of science. Yet, the new philosophers of physics have tended to go beyond van Fraassen's efforts to defeat the realist by denying that the realist's global game is one worth playing. The three most influential books in this vein have been Cartwright (1983) (on classical mechanics, thermodynamics, and electrodynamics), Friedman (1983) (on relativity), and Fine (1986a) (on quantum mechanics). Although their analyses differ according to the branches of physics they study, Cartwright, Friedman, and Fine share an aversion to any specifically philosophical accounts of why physics works as well as it does, instead preferring to appeal to features inherent in the practice of physicists, such as mathematical modeling and the reliable production of laboratory phenomena.

However, the realists have not exactly rolled over and played dead, though they have switched tactics. On the one hand, some realists have tried to reclaim the metaphysical high ground by arguing that their position must underwrite any coherent and comprehensive scientific worldview. In this context, realism becomes a cosmic materialism that is cashed out in various terms: evolutionary (e.g., Hooker 1987), neurological (e.g., Churchland 1979), or explicitly Marxist (e.g., Bhaskar 1979). On the other hand, more down-to-earth realists have lent philosophical respectability to the idea that something is real if it can be used to affect something else. The second half of Hacking (1983) is devoted to this topic, from which has emerged a distinguished set of case studies on experimental manipulation in physics that have tended to recreate the poles of the Laudan–Bloor debate. Not surprisingly, the authors of these studies have had top-flight training in physics: Franklin (1986) claims to have found philosophical canons of rationality in the conduct of physics experiments, while Pickering (1984) comes close to saying that reality is socially constructed in the lab, and Galison (1987) splits the difference between the two.

4. The Growth Areas: Biology and Cognitive Science

Over the past decade, the biggest shift of philosophical interest in science has been from physics to biology and cognitive science. (An interesting bibliographic feature is the virtual monopoly that MIT Press enjoys in both growth areas.) Unlike physics, which had attracted attention largely because of its exemplary status as a science, the reasons for younger philosophers especially moving to biology and cognitive science are quite diverse, reflecting a broader set of concerns than philosophers had previously considered. The point is perhaps clearest in the case of philosophy of biology, which has at least three distinct motivations.

The first motivation starts with the observation that "biology" is itself a rather loose term for a cluster of disciplines—ranging from population genetics to systems ecology—that are in varying states of methodological order. Thus, some philosophers (e.g., Rosenberg 1985) have made it their business to evaluate the scientific credentials of these disciplines and to suggest whether they would be improved by becoming more like physics.

Second, some philosophers engage in the more modest and meticulous task of conceptual analysis of key biological concepts, such as selection, fitness, and adaptation. Analysis of this sort serves a twofold function: on the one hand, it illuminates traditional philosophical problems of chance, causation, and explanation, which in the past have been dominated by examples from physics; on the other, it contributes to clarifying some of the impasses in recent debates within biology, such as the level at which natural selection operates (Brandon & Burian 1984) and the terms on which a unified biological science can be forged (Bechtel 1986). The major monograph in this category is Sober (1984).

The third motivation for philosophy of biology is the kind of "public outreach" with which Philip Kitcher has been most closely associated. When Creationism threatened to share space with Darwinism in high school biology textbooks, Kitcher (1982) shows precisely how Creationism fell short by standard philosophical accounts of the scientific method. More recently (and to much acclaim), Kitcher (1985) has provided a systematically cautious reading of the evolutionary evidence to discredit the claims of sociobiologists. Despite the polemical nature of both books, they demonstrate the potential of sophisticated philosophical analysis to make a difference in public debate. More strictly academic considerations of the limits that biology places on the human condition and human inquiry are now also the subject of two wide-ranging but accessible books: Ruse (1986) (a relatively temperate defense of sociobiology) and Levins and Lewontin (1985) (a quasi-Marxist attack by two leading theoretical biologists).

So far, we have considered the recent expansion of the philosophy of biology as literal extensions or applications of biological knowledge. But what about *metaphorical* extensions and applications? Since these figure periodically in this book, they are worth raising in connection with further developments in the attempt to naturalize epistemology and the philosophy of science.

Most contemporary philosophers receive their introduction to naturalism from the works of Willard Quine (esp. 1960, 1969), who basically was interested, much in the way John Dewey was, in using the variety of our experience to blur certain distinctions that the positivists thought were necessary for science to be understood as a rational enterprise. Of note here is the positivist distinction between the contexts of *discovery* and *justification*. The positivists held that the former was given to the vagaries of creative genius, while the latter could be rigorously formalized. Quine naturalistically translated these two contexts into questions of, respectively, perceptual psychology and sociology of knowledge. However, Quine was very much influenced by his Harvard colleague, B.F. Skinner, which gave his naturalism a strongly behaviorist cast. Thus, "perceptual psychology" is studied by looking at people's behavioral dispositions to link words to situations, and the "sociology of knowledge" is studied by looking at the communication patterns by which such dispositions are stabilized for a community.

Although these naturalistic translations could conceivably have led to the assimilation of epistemological problems by the social sciences, Quine's Skinnerian outlook led him to think in a more biological direction, even if it turned out only to be biological metaphors for psychosocial processes. Indeed, what is nowadays called "evolutionary epistemology" is precisely that. Skinner (1954, 1957) himself first indulged in this imaginative extension of biology when he conceived of all forms of learning as the "selective reinforcement" of "operant behavior," the former acting as a blind Darwinian environment that tests what the latter emits as hypothetical attempts at survival. In his later years, Karl Popper (1972) found this model an attractive way of capturing his falsificationist methodology, though the most sociologically realistic development of the "selective retention" model of knowledge growth belongs to Donald Campbell (1988). Trained as a behaviorist, Campbell sees the main normative problems facing the naturalized epistemologist as the construction of environments that select for features of hypotheses that are well-adapted to the nature of reality. If anything, Campbell has a tendency to reduce the generative side of the biological model (i.e., the analogue between genetic variation and hypothesis generation) to a "vicarious selection" process, whereby the scientific organism runs through in its mind the likely fates that various hypotheses would face if they were to be openly proposed.

Among philosophers of science, Stephen Toulmin's (1972) *Human Understanding* and David Hull's (1988) *Science as a Process* provide an instructive contrast in how the evolutionary model can be deployed. Although Toulmin alludes to the social and psychological character of science, as we shall see in Chapter Two, his major arguments for the evolutionary model are drawn almost entirely from the internal history of science, from which personalities and institutions are largely absent. Interestingly, it cannot be said that Toulmin lacks an understanding of the more traditionally "external" factors, for he was probably the first anti-positivist philosopher of science to translate his normative concerns into a policy arena, one in which he locked horns with both government economists and leaders of the scientific community (Toulmin 1968). For his part, Hull draws on his experience as the editor of a major journal in systematic zoology to portray the evolution of knowledge in that science as emerging out of the interpersonal conflicts of the participating scientists.

Although both Toulmin and Hull rely heavily on biological theory to guide their use of the evolutionary model, they take almost nothing from the social science literature when it comes to linking the model to the actual practices of scientists. Instead, they fall back on folk notions of why people do what they do. In the end, this is the ultimate weakness of evolutionary epistemology: Instead of using evolutionary mechanisms as suggestive analogues for genuine processes that a social science might seek to identify (if it hasn't already, see, e.g., Hawley 1950), these naturalists simply let an articulation of the analogy replace the necessary cross-disciplinary empirical work, which may serve to vindicate only some of the putative correspondences. Campbell is the exception that proves the rule here. However, it is unlikely that Campbell will serve as an exemplar for future research, until philosophers rid themselves of their distinctly "homeopathic" bias in constructing theories of science. The bias is captured by this question: *Why do theories of science always seem to have the same structure as the scientific theories that they hold in high esteem?* Specifically, isn't it a bit too coincidental that philosophies of science that tout evolutionary biology as the best theory around are themselves evolutionary theories?

Of course, the original internalist accounts of science portrayed scientific inquiry as closed under a small set of methodological principles, modeled on Newton's laws. And, if any discipline has capitalized on "the machine in the machine" image of science's place in the reality that it studies, it is certainly *cognitive science*. Perhaps the most striking difference between philosophical work in biology and cognitive science is the extent to which philosophers in the latter see themselves (often persuasively) not as mere underlaborers but as front-line contributors. No one has made more of his role as "cognitive scientist" in this sense than Jerry Fodor, one of Noam Chomsky's early collaborators in psycholinguistics (e.g., Fodor 1975), who, over the past ten years, has outlined several research programs, to which many have devoted their energies (e.g., Fodor 1983). Several reasons stand out for why philosophers exert such leadership in cognitive science. Aside from the relative infancy of the field (at most 30–40 years old), the interdisciplinary character of cognitive science creates the need for interlocutors and synthesizers, roles in which philosophers flourish. Certainly, the most lucid and engaging introductions to the field, Haugeland (1984) and Churchland (1984), have been written by philosophers. For a sense of the array of computer scientists, neuroscientists, psychologists, and linguists around whom philosophers must maneuver, one need only turn to a typical issue of *Behavioral and Brain Sciences*, which publishes programmatic statements followed by twenty or more commentaries. Finally, as Fodor (1981) himself has emphasized, cognitive science is still very much a study of the solitary thinker, be it human or machine, which puts the field squarely in the philosophical lineage of Descartes.

But is there really a "thinking substance," as Descartes thought, which warrants a strong separation of cognitive science from the rest of the natural sciences? Commonsense, or what is nowadays disparagingly called "folk psychology," would seem to support such a separation, but commonsense may itself be a false theory. Thus goes the debate that grips cognitive scientists today. Folk psychology has found its loudest defender in Fodor (1968), who definitely sees computers and chimpanzees as playing catch-up with the intricacies of human belief-fixation processes. What this view is supposed to insure, of course, is that philosophers—the people who invented the idea of commonsense as a theory in need of codification and obedience—will remain central to the future of cognitive science. However, one can also count on philosophers to offer arguments even for their own dispensability. The work of Stephen Stich (1983, 1990) is best understood in this way—Rorty (1979) prosthetically enhanced by a computer. It is a view that I take quite seriously in this book. More extreme is the work of the husband-and-wife team of Paul (1989) and Patricia (1986) Churchland, who presume that it has already been shown that folk psychology is false, and that cognition is best seen as a branch of neuroscience. Studiously occupying the middle ground in this debate is Daniel Dennett (1978, 1987a), whose witty books, *Brainstorms* and *The Intentional Stance*, offer a defense of the instrumental value of treating people as though they had rational minds even if it turns out (as he imagines it will) that minds are just complicated bits of matter whose overall efficacy can be judged only in the fullness of evolutionary time.

The entanglement of empirical and conceptual issues that arise from these positions in cognitive science has itself drawn considerable attention from philosophers of science. Cummins (1983) argues that progress in psychology has been hampered by psychologists operating with simple-minded notions of physical explanation drawn from logical positivism. He suggests that cognitive science has put psychology on a more secure scientific footing with the introduction of the computer program, which formalizes the idea of explaining mental states in terms of their "function" in a system of beliefs and desires. Wilkes (1988) looks forward to the day when philosophical speculation about the mind will be constrained, though not determined (*pace* Churchlands), by certain curious facts about our neurophysiology, facts that challenge the folk assumptions underlying even some of the crazier thought experiments philosophers such as the ones mentioned above (and especially Derek Parfit) are prone to use to stake out their positions.

One of the most fascinating features of the folk psychology debates is that the arguments for restricting the scope of a "thinking substance" to ordinary human mental faculties appear strongest when posed in a relatively popular, untechnical way. Searle (1984) and Hubert and Stuart Dreyfus (1986) are two of the clearest cases in point. However, it is by no means clear that restricting the scope of "thinking substance" to human beings is in the best interest of most flesh-and-blood people. After all, when "human beings" are used as the standard of cognitive competence, not just any old human being sets the standard, but a human who has been acculturated in a manner that is "appropriate" to the skill or task under study. In other words, what would normally be seen as the product of a privileged form of socialization gets interpretively built into the hardware of "expert" human beings, against whom both other humans and machines must measure up (Bloomfield 1987, Collins 1990). Privileging of this sort is less likely to occur if the exemplars of cognition are taken to be, not a particular class of human reasoners, but rather a generic, non-human class of "cognizers" that includes certain machines and animals, plus all humans, among its members. Indeed, humans may not even hold pride of place in this class. Although such a view may seem radical, it is none other than the "computationalist orthodoxy" espoused most systematically by Zenon Pylyshyn (1984) in *Computation and Cognition*. In this book, we take Pylyshyn's view in the direction of "android epistemology, " or the "The Robot's

Dilemma," to quote the title of a recent anthology he edited (Pylyshyn 1987).

If a "cognizer" is nothing but a set of materially diverse entities that instantiate certain formal patterns of activity that are recognizably "thoughtful," then why must we even restrict the class of cognizers to individuals? Herbert Simon, the Nobel Prize winning economist (trained in political science), was one of the first to develop "general problem solving" machines in the late 1950s. He has all along maintained that computers may model societal as well as individual processes, all of them requiring "artificial" intelligence in the strict sense of having to devise means of overcoming an environmental resistance to reaching a self-generated goal (Simon 1981). Thus, cognizers may include entire business firms as well as chessplayers, individual scientists as well as the entire history of science rationally reconstructed. This last project has been the latest source of inspiration for Simon and many of his students (Langley et al. 1987, Shrager & Langley 1990). Indeed, the best cases of interdisciplinary activity in cognitive science—involving philosophers, psychologists, and computer scientists—have been in the area of scientific reasoning. Holland et al.'s (1986) *Induction* is the most celebrated case in point. Scientific reasoning has also been the site of the most celebrated interdisciplinary controversy between cognitive scientists and sociologists of science over who is the real heir-apparent to epistemology. The November 1989 issue of *Social Studies of Science* centers on a spirited piece by Peter Slezak, the head of Australia's first Cognitive Science Unit (which is located in an STS Department!). Slezak argues that machines such as Simon's refute the Strong Programme claim that scientific discovery always requires a social context. Would it be too much to add that Slezak did his doctoral dissertation on Descartes?

5. An Itinerary for the Nineties: Does Science Compute?

Tempting though it may be to delve into the dynamics of the Slezak controversy, which eventually enveloped many of the people discussed in the pages of this book (including Herbert Simon himself), I will refer the reader to the analysis provided in my latest book (Fuller 1992a, ch. 5) and turn instead to a discussion of Tweney (1990), the closing critical chapter of Shrager and Langley (1990). The issues that Tweney raises in his "Five Questions for Computationalists" center on the prospects for integrating the work of historians, experimenters, ethnographers, simulators, philosophers, and other social scientists into one "cognitive science of science" (cf. Downes 1990). Whatever conceptual tensions already exist among these methodologies are typically heightened when the object of concern is science itself. For openers: To what extent do computational models of scientific reasoning need to be responsive to the constraints on human performance that experimental psychologists have repeatedly shown? Pylyshyn is among only a bare handful of psychologists (John Anderson and Philip Johnson-Laird are two others) who have tried to articulate the terms in which the findings of human- and computer-based research can be integrated into a unified theory of cognition. Interestingly, they typically ease their burden by downplaying or reinterpreting the phenomenology of human experience, which tends to cast the data of sensory qualia, mental imagery, and skilled practice in ways that *prima facie* elude the capabilities of computer models. A common tactic is to claim that mental imagery, say, efficiently represents to the cognizer information that is really stored propositionally (e.g., the image is an unarticulated analogy) but would take up too much processing space if it were consciously available to the cognizer as a set of propositions. Although I myself am drawn to such an account, Tweney will not let the burden fall so easily.

Tweney appears periodically in the pages of this book because he has worked in virtually all of the naturalized methods at one point or another. In the 1970s, Tweney and his colleagues at Bowling Green State University (Ohio) did important experimental work that extended Tversky and Kahneman's research on cognitive biases and limitations to the study of scientists (or students given scientific reasoning tasks). The Bowling Green group found that education in science—even in statistics—did not seem to counteract the effects that Tversky and Kahneman originally found (cf. Tweney et al. 1981, esp. parts 4, 6). As many of the experimental tasks were designed to test canonical models of rationality, the general conclusion reached was that these models have little purchase on our psychological makeup. From there Tversky and Kahneman attempted to recast the liabilities they uncovered as "heuristics" that work well for a limited range of cases but not for all. For his part, Tweney refused to accept the assumption that scientists were as deeply flawed as the experimental evidence suggested. Instead, he was inclined to believe that the native intelligence of scientists lay in the construction and maintenance of what may be called "intellectual ecologies." These ecologies could not be simulated in the psychologist's lab on the basis of a few experimental protocols, but rather require co-workers, technical apparatus, visual representations, and texts—especially notebooks, which Tweney characterizes as "external memory stores," specially crafted structures (often with their own idiolects) designed to prime the scientist's cognitive powers.

Over the last ten years, Tweney has been a major contributor to the cognitive history of science, largely by reconstructing Michael Faraday's intellectual ecology from archival materials. All along, Tweney has portrayed himself as adhering to the division of labor originally prescribed by Wilhelm Wundt at the scientific dawn of psychology. Wundt argued that laboratory experiments were appropriate to the study of the material interface between the organism and the external world, namely, sensation. However, higher thought processes involve culturally mediated extensions of experience— especially language—which can be fully fathomed only by more hermeneutical methods. Thus, Wundt believed that psychology unified the sciences of "nature" and of "spirit." But where would a latter-day Wundtian, like Tweney, place Shrager and Langley's collection of computational models? It is in this spirit that Tweney raises his five questions:

(1) Can the model handle higher-order heuristics?
(2) Can the model handle chancy interactions?
(3) Can the model handle visual imagery?
(4) Can the model handle a large knowledge base?
(5) Can the model distrust data?

In their introduction, the editors anticipate some of Tweney's concerns by drawing a useful distinction between the *embodiment* and the *embeddedness* of scientific cognition. Consider the individual scientist. Even if she is in the company of other scientists who are willing and able to listen to what she has to say, the scientist nevertheless has to face "the enemy within"—the various ingrained habits, motivational biases, and processing limitations that channel reasoning in ways unbeknownst to the reasoner herself. This point highlights the fact that reasoning is never under the full control of the reasoner, but is rather subject to constraints that stem from the particular body in which the reasoner happens to find herself. But even if our scientist were able to take steps to minimize the impact that embodiment has on the course of her reasoning—say, by deploying a battery of diagnostic and prosthetic devices—she would still be left with the task of embedding her thought in a social context that enables her to address the relevant scientific audience. Once again, success at this task is not guaranteed. Differences in background assumptions and research interests—not to mention issues pertaining to the embodiment of the individual members of the audience—can impede communication. In the long term, these problems may be compounded by the unanticipated use of the scientist's work by opportunistic third parties who end up redirecting the course of research away from the scientist's original agenda. The issue here is not whether embodiment and embeddedness are sufficiently serious issues to worry the computationalist—for they most certainly are—but whether these issues can be adequately handled by simply adding background constraints to the core Cartesian cognizer that the computer is normally taken to represent (cf. Fodor 1981, pp. 100–126). Below I suggest what an alternative vision might entail.

Tweney quickly gets to the heart of the matter by identifying himself as a "consumer" of computational models of scientific reasoning. This triggers a host of questions: Who is in the market for a computerized scientist? Who will (fail to) be persuaded that the computer is a scientist? In its canonical form, the Turing Test fails because it presumes the naive sociology to which these questions are addressed, namely, the belief that *any* human could be used to determine whether a given computer displays intelligence. But it may be that Tweney the cognitive historian would get little out of machines that fail to reenact the thought processes of human scientists, whereas Tweney the cognitive psychologist might prefer the advice offered by the very same machines to that of human colleagues when it comes to deciding on which experimental hypothesis to pursue. Moreover, if the computer has got a good track record in recommending hypotheses, why should the practicing scientist worry about whether the computer arrives at its recommendations in the same way she does? Not only is such a requirement not normally placed on one's human colleagues, but it may not even make good ontological sense to do so.

Why should the individual—human or computer—be seen as an epistemologically self-contained unit capable of supplying good reasons for any conclusion it reaches? Instead, why not cast the individual as a proper part or stage in a collective process, whose overall operation determines post facto the treatment given to the input contributed by any of its constituents? From this alternative viewpoint, the computer will have been proven adequate in its reasoning if scientists continue to seek its advice in the future as part of the normal course of their activities. Although Tweney alludes to such a perspective in addressing question (3), he does not take it seriously enough, as suggested by the evidence he musters for the claim that Faraday operated with a much larger knowledge base than any computer can currently handle. The evidence is the detailed character of Faraday's notebooks, which Tweney treats as an elaborate mnemonic for retrieving knowledge that Faraday already possessed. However, the alternative viewpoint would suggest that Faraday constructs his knowledge of electromagnetism de novo in the course of interacting with the notebooks. In other words, Faraday himself is merely one part of a cognitive system that constitutes "Faraday's knowledge." It may or may not be the case that a computer can model that system, but if it can, it may well be modeling a knowledge base that extends beyond the cognitive capacity of a single human being.

Tweney casts questions (1) and (2) in terms of the temporal and spatial arrangement of the computer's learning environment, but the issues that they raise fall under the more general heading of scarcity management. Consider the more general problem first. A firm has limited resources that it can invest in a variety of ways. How does it determine the best course of action? Economists have long realized that it all depends on the firm's place in the market: Does it enjoy a monopoly or does it see its market share declining? The answer turns on the narrative that the firm tells about itself, which will, in turn, lead it to be either inclined or averse to certain levels of risk. So too with the scientist who must decide whether to continue pursuing a line of inquiry that has so far yielded mixed results or to strike out in a new direction entirely. Process and opportunity costs need to be calculated. The former set of costs involve how tomorrow's inquiries are likely to be shaped by today's choices, whereas the latter set refer to alternative lines of inquiries that are permanently preempted by today's choices. Clearly, these calculations cannot be made without the scientist having a robust sense of their place in the history of inquiry.

As for the spatial arrangement of the learning environment, Tweney's point here is that creativity may be fostered by allowing a certain amount of interaction to take place between the distinct problems on which a scientist is currently working. Yet, clearly, too much cross-fertilization can breed spurious connections and overall confusion. This issue frequently arises in discussions of the institutional imperative to foster interdisciplinary research. The claim that maintaining disciplinary boundaries is better than having none at all is completely compatible with the claim that it is better for those boundaries to be permeable rather than rigid. The relevant economic analogue to gauging the appropriate level of cognitive cross-fertilization, or interdisciplinary permeability, is the problem of locating firms so as to enable an optimal level of competition in the marketplace. On the one hand, economists know that there should not be such a high density of competing firms in one location that it becomes impossible for the consumer to make an intelligent choice between them. A superabundance of access to products is no access at all. But on the other hand, too few firms can spawn monopolistic practices, thereby stifling competition and the drive to product innovation.

How far can the analogy between economics and cognition be pushed before it simply becomes absurd? I would say quite far, probably much farther than Tweney would countenance. The circumstantial evidence for this claim is certainly suggestive. As philosophers know, there is a long-standing metaphysical tradition of nominalism that explains the existence of universals in terms of our finite ability to hold individual differences in mind. And, of course, Herbert Simon's own career is testimony to the ease with which one can think about the scarcity of material and cognitive resources by using many of the same tools. Perhaps there is even a cognitive equivalent to Mandeville's maxim that "private vices make for public benefits": to wit, *local errors make for global principles*. For instance, in addressing question (5), Tweney confronts computationalists with the hoary challenge of programming a computer that knows when to break the rules of method so as to seize upon a genuine discovery. But why presume that the machine is stocked with rules in search of appropriate ceteris paribus clauses? Instead, maybe we should imagine that the machine is supplied simply with rules that it fails to apply reliably. Perhaps the computer would, indeed, be too inferentially conservative to be scientifically interesting, if it were capable of following the rules "mechanically." But luckily, *it's not that smart*, and hence it is receptive to results that stray from the strictures of method. In other words, the trick of learning from learning may be to figure out which cognitive imperfections need perfecting and which ones are good enough just as they are.

6. The New Wave: Metascience

If one were confined to the world's premier philosophy of science journals, *Philosophy of Science* and *British Journal for the Philosophy of Science*, one might start to worry that the standards of normal science in particular fields have overtaken more critical, strictly "philosophical" standards. (A rough test for the amount of philosophy in a philosophy of science article is the number of claims it contains that challenge the conventional wisdom of the field under study.) In response to this worry, and wider cultural trends, a growing number of non-philosophers and philosophers outside the mainstream have returned to the more general normative issues about the conduct of inquiry that fueled the academic imagination during the Kuhn debates.

These works, which fall under the rubric of "metascience," are a diverse bunch, often willfully blurring the lines that have traditionally separated description from prescription. Taken as a group, "metascientists" tend to be more interested than most of the previously cited authors in reforming the practices of science, though the motivation for such reforms vary enormously from purely internal considerations to more globally societal ones. Although no one journal deals adequately with the issues raised in all of these works, the journal *Social Epistemology* has endeavored to address each of them at some point.

Twentieth century "Continental" European theorists of science differ from their "Anglo-American" counterparts in how they focus their normative concerns. As we have seen up to this point, English-speaking philosophers neatly dissociate science from technology, agreeing that the ends of science are noble, but the best means for their pursuit remains the subject of "methodological" dispute. By contrast, Continental theorists tend to collapse science and technology into the behemoth, "technoscience," and, consequently, take the very ends of science to be an eminently contestable issue. This has led to several radical critiques of science, typically inspired by either Heidegger or Marx. (We already had a taste of these critiques in our brief discussion of technology.) Both Heideggerians and Marxists claim that the authority of science in modern society rests on certain myths of rationality and realism that have become associated with "the scientific method." Their common antidote is to understand what scientists do in their workplaces and how their practice has subtly colonized everyday life. But there the resemblance ends.

The Marxists are generally pro-science, but they want science placed squarely in the service of society (e.g., Schaefer 1984). They tend to treat science much as they do capitalism—when they distinguish them at all (admittedly, a trying task in this age of "Big Science"). Thus, Marxists fault science, less as a mode of knowledge production than as the set of social relations that science has historically supported, relations which have restricted the emancipatory potential of scientific knowledge. Telling in this regard is Marxism's relative indifference to the ecological consequences of Big Science (an exception is Aronowitz 1988). Here, by contrast, the Heideggerian reaction to science comes to the fore. Heideggerians tend to distrust the instrumental value of science, believing that science only serves to alienate humanity from its natural form of existence. Although originally trained in the Heideggerian way, Joseph Rouse (1987) has masterfully combined significant elements of the Marxist critique, feminism, and post-positivist philosophy of science, to forge a "postmodern political philosophy of science." In a nutshell, "postmod-

ernism" is the philosophical attitude whereby one converts into a
virtue everything that a positivist would regard as an adversity. This
attitude applies, in particular, to the current disunity of the sciences.
The first postmodern introductory textbook in the philosophy of
science is Ormiston and Sassower (1989). D'Amico (1989) usefully
relates these newfangled concerns to the fundamental problem of
humanistic knowledge, namely how is transhistorical knowledge
possible in a historically situated world.

If one can speak of a "grassroots" in metascience, it would have to
be the Rhetoric of Inquiry, a loose confederation of scholars who study
the means by which arguments persuade in their respective disciplines.
(University of Wisconsin Press is their publisher of choice.) Most of
these scholars came to rhetoric after having distinguished themselves
as practitioners of one of the human sciences. They are joined by a
concern that disciplinary specialization impedes the growth of
knowledge by restricting communication and accountability, which
has dire consequences once these restrictions become marks of
authority in the public sphere (Nelson et al. 1987). Thus, the Rhetoric
of Inquiry has contributed substantially to the debate over the role of
experts in a democracy. The argumentation scholar, Charles Willard
(e.g., 1983, 1992) is the leading figure in this area. I try my own hand
at this field in my latest book, *Philosophy, Rhetoric, and the End of
Knowledge* (Fuller 1992a), which preaches interdisciplinarity as an
ideology with implications both in and out of the academy. A legion of
allies for the Rhetoric of Inquiry can be found among technical writing
instructors, who have made concrete proposals for academics improving their cross-disciplinary and extra-academic communication skills.
For example, Charles Bazerman (1987) has studied the rhetorical
consequences of the variety of journal formats ranging from physics to
psychology and English.

The two disciplines in which the Rhetoric of Inquiry has made
the most impact are the ones whose scientific credentials have been
most loudly trumpeted and contested: economics and psychology. In
economics, the issue turns on the relevance of often abstruse
mathematical expression to the assessment of theoretical claims. A
mathematically accomplished economist in his own right, Donald
McCloskey (1985) is often said to have founded the Rhetoric of
Inquiry movement, when he declared that economists often revert to
mathematics in order to shield their claims from public scrutiny.
However, since then, Philip Mirowski (1989) has concluded that
McCloskey's diagnosis only scratches the surface, as it underestimates
the extent to which this mathematical rhetoric has served to shape the
way in which economists think about their research. Mirowski goes so

far as to claim that economists are captive to equilibrium models derived from now defunct branches of physics. Scrap the mathematical rhetoric and its physicalist associations, says Mirowski, and the scientific illegitimacy of economics will come into full view.

While metascientific critiques in economics are largely performed by accomplished practitioners of the techniques being criticized, in psychology the situation is somewhat different. Here a long-standing alternative tradition, especially in social psychology, has opposed the positivistic assumptions that are seen as underwriting the experimental method. This tradition, still more influential in Europe (including Britain) than in the USA, is aligned with the "kinder-and-gentler" Oxford realists discussed earlier. However, a more radical "constructivist" strain among those trained principally in psychology has been generated by Kenneth Gergen (1983) and John Shotter (1984). They have argued for a more ethno-historical approach to the human sciences that is in the spirit of much of the empirical work in Science & Technology Studies (cf. Gergen 1985). Historians have aided in the constructivist cause by showing famous instances in which psychological experiments were more successful at suppressing latent power relations between experimenters and subjects than at revealing any regularities in human behavior beyond the laboratory setting (Morawski 1988, Danziger 1990). These historians suppose that a deconstruction of the lab's power relations is sufficient to undermine whatever external validity claims the experiments claim to have. Greenwood (1989) is a philosophical attempt to mediate the realist and constructivist metascientists.

7. Feminism: The Final Frontier?

The most comprehensively radical critiques of science available today are due largely to *feminists*, who have proceeded independently of—and often with hostility from—mainstream philosophy and sociology of science. Yet, "feminism" is itself just as internally divided as the philosophy and sociology of science.

For newcomers, Sandra Harding's (1986) award-winning *The Science Question in Feminism* draws the most connections to mainstream philosophy of science (esp. Kuhn) in arguing for a multiperspectival understanding of science as the way to dispel myths of realism and rationality that perpetuate the domination of masculine values. Harding's (1991) follow-up book, *Whose Science? Whose Knowledge?*, includes a critique of current sociologies of science, while foregrounding the politically and epistemologically sensitive issue of the differences between white Western middle-class feminist approaches to science and approaches that are currently being pursued by various combinations of non-White, non-Western, lower-class men and women. For example, while it may be liberating for someone in Europe or Anglo-America to try to subvert the authority of science by showing that the effects of science vary across contexts, uttering such an argument in parts of Africa, Asia, or Latin America would only serve to reinforce the dominant classes by discouraging the mass introduction of labor-saving technologies. Harding realizes that the normative agenda of tomorrow's science will need to address problems of that sort. In addition, Harding has also compiled two useful anthologies that address feminist issues at finer grained levels of scientific practice and from more radical viewpoints: Harding and O'Barr (1987) centers on the life sciences, while Harding (1987) focuses on the social sciences.

Feminists have generally been more vigilant than most philosophers in not letting multi-perspectivalism slip into a "judgmental relativism" that accepts what anyone does on their own terms. Harding's way of avoiding this problem is by adopting a "standpoint epistemology," which argues (à la Lukacs vis-à-vis the proletariat) that excluded or dominated classes may have a more objective understanding of the social order because they lack a clear vested interest in its maintenance. A more conservative way out of relativism is Helen Longino's (1990) "feminist empiricism," which argues that the British empiricist tradition stemming from Locke and Hume has operated with an impoverished sense of "sensory experience," ultimately reducible (so it would seem) to one man's pair of eyes. The remedy is to portray science as involving an entire community of fully embodied inquirers. Longino herself, however, often does not distinguish feminist from sociological or generally pragmatist contributions to her brand of empiricism. In fact, she often makes it appear that feminist critiques can be applied only to contested areas in the social and biological sciences. This, of course, goes against the sorts of global feminist methodologies championed by, say, Evelyn Fox Keller (1985), which range over the physical and biological sciences, as well as history, philosophy and sociology of science. Taking a broadly psychodynamic approach, Keller argues that the scientist's very sense of identity (regardless of gender or discipline) has too often depended on internalizing strict subject/object dichotomies that are the hallmark of masculinist thinking. Keller unearths these dichotomies from the etymologies of key words in scientific discourse, which presumably impart unconscious associations to their users. Unlike Longino and perhaps more so than Harding, Keller envisages feminism transforming not only the aims of science but its actual daily practice.

The feminist critic of science to receive the most attention in recent years is perhaps the most radical of the ones we have so far considered, Donna Haraway. Haraway (1989) locates the masculinist bias in the methods used to interpret primate behavior—in the wild, in captivity, and in the public forums (ranging from philosophy to advertising) where claims about primates figure prominently. She argues that the nature/culture distinction results from forms of human domination being projected onto the animal kingdom. What is perhaps most distinctive about Haraway's work, which has earned her many emulators, is that it reveals the tractability of science to the research methods of "cultural studies," a curious attempt to turn postmodernism into normal science. Nevertheless, cultural studies incorporate areas of the humanities left out by HPS and not yet fully assimilated within STS: deconstruction, semiotics, and other approaches to media criticism, which serve, in turn, to blur the boundary between scientific and popular culture. In her most recent work, Haraway (1991) has used the science-fiction image of the "cyborg," the human–animal–computer hybrid, to signal the need for a new metaphysics in a world that is quickly developing gene-based technologies and computational neural networks. Together they have blurred such traditional ontological boundaries as organic/mechanical and innate/environmental. But Haraway's "cyborg manifesto" is aimed at feminists just as much as philosophers, in that feminists have often shied away from talking about the "biology" of gender relations for fear of slipping into a reductionism or essentialism that makes the difference between men and women fixed and unalterable. By observing that biology need not have these associations any longer, Haraway hopes to "re-enchant" science with a newfound emancipatory potential.

CHAPTER TWO

Mythical Naturalism and Anemic Normativism: A Look at the Status Quo

1. The Mythical Status of the Internal History of Science, or Why the Philosophy of Science Is Suffering an Identity Crisis

Can a discipline suffer an identity crisis? Presumably, if it either loses its subject matter or discovers that it has lacked one all along. Richard Rorty (1979) took the former position to criticize all of philosophy in *Philosophy and the Mirror of Nature*. To compensate for the restricted scope of my critique, I am resorting to the more radical latter position. I aim to show that, strictly speaking, philosophy of science has an imaginary locus of inquiry, known as the *internal history of science*. Its main historiographical assumption is that there is a natural trajectory to the development of science when it is regarded as a knowledge producing activity. Internalists divide over even this simple definition, especially over the identity of the relevant "regarders." Some, like Imre Lakatos (1981), believe that only the philosophically informed historian, by ridding the history of science of the social impurities that typically vitiate scientific judgment, can see the natural trajectory of knowledge production. This, in turn, gives the Lakatosian a privileged perspective from which to offer advice on the future conduct of science. Others, like Dudley Shapere (1987), believe that scientists themselves come gradually to perceive this trajectory over the course of history as they "learn to learn" about the world. Shapere's philosophically informed historian is then directed to identify the emerging patterns of knowledge production across the disciplines. Most internalists seem to fall somewhere between Lakatos and Shapere, though my later discussion of Larry Laudan's (1984, 1987) recent work will show that the middle ground is much more unstable than it first may appear.

The internal history of science is not just another private philosophical fantasy but one that has seduced the imaginations of other humanists and social scientists with an interest in the nature of science. These inquirers have, for the most part, uncritically accepted the burden of defending an "external" history of science, even though little more is conveyed by the expression than a general opposition to internalist history. In practice, however, an externalist is anyone who believes that science, like any other human activity, cannot be discussed intelligibly unless it is treated as something that is essentially embodied, where "embodied" means being bounded in space and time. For example, when psychologists appeal to human cognitive limitations or sociologists appeal to local interests in order to explain the course that the history of science has taken, they are treating science as essentially embodied. Despite the natural advantage of the externalist program from the standpoint of a unified science of human behavior, externalists have undercut this advantage by acquiescing to the image of history that internalism implies, an image borrowed from the principle of inertia in Newtonian physics, which postulates that bodies continue in a regular motion unless subject to interference. The "body" in this case is what philosophers call "rational methodology," that is, reliable means for attaining some end, normally called "truth." Given this image, externalists seem to be concerned only with what draws the scientist away from the course that rational methodology projects, a decidedly second-class enterprise.

As we are beginning to see, an entire rhetoric accompanies the internal–external distinction and the debates that have since engulfed philosophers, historians, sociologists, and, most recently, psychologists. A quick guide will prove useful for understanding what follows. Since the existence of an internal history of science looks more plausible with ever more abstract characterizations of the actual history, there is considerable ambiguity about the referential function of such key internalist terms as "science," "rationality," "truth," "autonomous," "epistemic," and "cognitive." But here are some rules of thumb. "Science" tends to be the name given to the activity to which the other five terms pertain to the greatest extent, usually as determined by today's philosophically informed historian. In practice, this means that physics is virtually everyone's paradigm case of a science. "Rationality" and "truth" are more often described in functional terms than in truly substantive ones. Thus, "rationality" is to "truth" as "means" is to "end." To say anything more substantive would be to open the Pandora's box of positions in the philosophy of science.

However, one implication of defining rationality and truth in purely functional terms is that the sorts of acts that might count as violations, or dysfunctions, of rationality or truth are always in the back of the internalist's mind. Indeed, this point serves to motivate the idea that science is a distinctive enterprise, one which cannot merely adapt to its ambient social setting, but which must often resist the call of the social. When scientists engage in this resistance successfully, they are "autonomous," a term with unusually strong psychologistic overtones by internalist standards. Externalists have something to study (so internalists claim) precisely because science is frequently dysfunctional, largely because scientists prove not to be autonomous. Finally, "epistemic" and "cognitive" are increasingly used as synonyms, though typically the former refers to science as a rationally organized body of knowledge (as in the German *Wissen*), whereas the latter refers to science as a rationally directed way of knowing (as in the German *Erkennen*). The two terms often appear as modifying "factors," when activities that make positive contributions to the production of knowledge are being discussed (for a history, cf. Graumann 1988). That they are used almost interchangeably reflects the extent to which internalists presuppose that there is an inherent link between the quality of knowledge processes (cf. cognitive) and the quality of knowledge products (cf. epistemic).

In addition to stacking the terms of the debate in her own favor, the internalist prevents externalist encroachment by deftly moving between historical modalities. An especially agile slide is from what *did* happen to what *ought to have* happened, via what *would have* happened—if certain ideal conditions had been historically realized. Indeed, an internalist might go so far as to blame, not the historical agents, but the historian herself for being unable to abstract the essential course of scientific progress from the surfeit of data that confronts her. However, this aggressive methodological posture conceals a remarkable weakness on the philosopher's part. The internal history of science is supposed to describe those episodes in which science has been done "by the rules," that is, in ways that philosophers have reason to believe will provide most direct access to truth. As it turns out, there are rather few of these episodes, which suspiciously correspond to the most famous incidents in the history of science (Newton, Lavoisier, Darwin, Einstein), which even more suspiciously are presumed to play a significant role in explaining the less exemplary episodes (e.g.,, the scientists in these episodes were merely imitating, anticipating, or completing the work of Newton, Lavoisier, Darwin, Einstein). We will see early in this chapter that such a view of the history of science cannot withstand much scrutiny.

Not only do internalists slide between value and causal significance, but they also presume a curious relation between *normative* and *descriptive* accounts of science. Since so few episodes conform to her preferred rules, the internalist is forced to say that most of the history ought to have been other than it was, which, as far as the entire history of science is concerned, means that she is offering a largely normative account. The complementarity suggested here between descriptive and normative adequacy is generally taken for granted in contemporary philosophy of science. For example, when Karl Popper (1972) began to think that falsification increased the reproductive capacity of the human species, he started to deny that falsification was "merely" a normative account of science but argued rather that it was part of the "situational logic" underlying all evolutionary developments. Of course, it is entirely possible for a philosopher to be blessed with most of the history of science conforming to her normative account. Larry Laudan (1977, ch. 5), for one, has explicitly aspired to be such a philosopher. Nevertheless, the normative aspect of the internal history of science tends to be asserted only once its descriptive inadequacy has become clear. Indeed, the greater the discrepancy between the internalist's norms and the historical record, the more likely that the discrepancy will be blamed on the scientists' failure to live up to the norms. In short, norms become the "higher ground" to which the philosopher's falsified account retreats!

This glimpse into how the philosopher manages not only to salvage but also to strengthen her normative account in the face of resistance from the history of science is just one of many rhetorical strategies that internalists have developed in order to keep the burden of proof firmly planted on the externalist's shoulders. I will simply comment on three of these strategies and leave it to the reader to spot them as she reads through the arguments addressed in this book and in many of the works that I cite. The reader will notice that a general feature of these strategies is that they involve an implicit double standard.

The first strategy is to circumscribe the sphere of the normative so that the philosopher presents herself as *evaluating*, but not actually *directing*, the development of science. This gives the philosophy of science its apolitical sense of critique, which is reminiscent of evaluation in aesthetics. The strategy can be seen at work in the preference that philosophers have shown for talking about what sort of research should have been pursued at a certain point in history (when nothing can now be done to change what happened) over what sort should be pursued in the future (when something can be done to change what would otherwise most likely happen). At face value, the past-to-present and present-to-future modes of science evaluation are susceptible to empirical treatment to roughly the same degree (e.g., both involve reasoning with counterfactuals [cf. Elster 1978, 1983]), yet since the futuristic mode invites explicit policy direction in a way that the historical mode obviously cannot, the futuristic mode has been shunned by philosophers.

But what, then, does the philosopher gain by restricting her normative focus to evaluating? By foreclosing the option of intervening in policy matters, the philosopher is also discouraged from engaging in a causal analysis of her own value judgments, which is to say, her appraisals of good and bad science are never subject to feasibility constraints. An internalist like Lakatos can declare that the history of science could have proceeded much more rationally by taking a course entirely different from the one it actually took, without his contemplating the costs and benefits that would likely have been incurred by the alternative trajectory. In short, an aversion to policy blinds the philosopher to the *fact-laden character of values*. What appears to be the optimally rational move from the standpoint of abstract philosophical criteria could turn out to be an implementation nightmare, which on any non-philosopher's theory would count decisively against the rationality of such a move. But, of course, to court facts in this way is to conjure up the image of science as an embodied activity, which would, in turn, offer the externalist too much dialectical leverage.

This brings us to the second rhetorical strategy by which the internalist advantage is maintained, namely, by manipulating the level of abstraction at which science is described. As I shall show below, whenever a concrete example of autonomous science is sought, social, political, economic, and psychological factors are immediately brought to the fore. However, philosophers nimbly deflect the full force of these moves by straddling the two extreme attitudes represented by Lakatos and Shapere. On the one hand, internalism can be defended (à la Lakatos) by arguing that a preferred model of scientific rationality can be instantiated by an indefinite number of sociologies, given that a methodology like falsificationism does not imply a specific way of organizing knowledge production. For example, it is open as to whether the "conjecturing" and "refuting" functions of falsification are to be seen as two cognitive processes operating in one scientist, in two distinct groups of scientists, in the same group of scientists, each of whose members is her own conjecturer and someone else's refuter—and the list of possible sociologies goes on. On the other hand, internalism can be defended (à la Shapere) by arguing that the preferred model of scientific rationality is simply the one that has in fact evolved during the history of science. The philosopher pitches her description so abstractly because she wants to concentrate exclusively on those features of the actual situation that have contributed to the growth of knowledge. In that case, to bring in the extra sociological detail would simply be redundant, if not potentially misleading to the naive historian.

What is missing between the Platonism of Lakatos and the Hegelianism of Shapere is the conceptual space that the externalist aspires to stake out, namely, the empirical preconditions, or "feasibility constraints," for the various models of scientific rationality. *Contra* Lakatos, not every social arrangement could (even in principle) support a preferred model of rationality; but *contra* Shapere, the current social arrangement is not necessary to the growth of knowledge and in fact could probably be improved upon. In its role as the normative wing of externalist history of science, social epistemology proposes to explore the implied intermediate possibilities (cf. Meja & Stehr 1988; Schmaus 1988). Even a move as scientifically basic as distinguishing "theory" from "observation" is not innocent of implicit sociological considerations. At the very least, such a distinction normally presupposes a division of labor between observers and theorists whereby the theorists ordinarily trust the reports of observers, whose skill was developed independently of any training in the theories judged by those reports; yet the theorist's trust in the observer's reports does not extend to determining what the theorist must do in light of such reports; and so forth.

The third, and final, rhetorical strategy is undoubtedly the subtlest, as it pertains to what counts as the "content" of science. I will shortly sort through the philosophical vexations surrounding this notion, but for now it will suffice to enumerate the sorts of things that the internalist wants to include and exclude by raising the specter of content. I imagine that a reader not already familiar with recent philosophy of science will find this enumeration somewhat undermotivated. For example, the private thoughts (what used to be called "the context of discovery") of the scientists are generally excluded, unless (as in the case of Faraday, Darwin, and other meticulous diarists) they are written so as to be accessible to methodological (or "cognitive") analysis. The public expression of science in journals and books is generally included, insofar as this reflects what the internalist can recognize as "the logic of justification." Any feature of the journals and books that seems to be driven by the "mere" pragmatics of scientific writing, such as article formats, length, and citation conventions is excluded.

Given that the transmission of knowledge depends on these pragmatic features, the internal history of science finds itself in the peculiar position of being a theory of scientific change that lacks any account of science as a process (cf. Hull 1988). Moreover, while internalists are not oblivious to the fact that what one says (in print) and what one does (in the lab) often diverge in significant ways, they are not thereby moved to examine the nature of these discrepancies. In fact, field studies of what scientists do in their work sites are generally regarded as centrally in the purview of externalist sociologists. Indeed, Maurice Finocchiaro (1973, chs. 16–17) has observed that the mark of the internalist is to ignore a basic canon of the historical method, a canon that would have utterances be treated as just more events to be explained, rather than as transparent accounts of other events. Thus, whereas an internal historian typically regards a scientist's statement of method as a description of her actual practice, it would be more historiographically sound to regard the scientist's statement as a retrospective rationalization of whatever she happened to have done, which would then need to be determined by other evidence. Why does the internalist blithely privilege what scientists say over what they do? A clue to the answer lies in the internalist's residual Platonism, which interprets the surface of science's own record-keeping as deep because of the close resemblance that the trail of scientific journals and books bears to something that a Platonist would readily find significant, namely, the dialectical development of a system of propositions. Indeed, the extent to which the scientists' background deeds deviate from their up-front words perversely feeds the Platonic impulse by suggesting that the inability of scientists to act as they should does not prevent them from knowing, and hence saying, how they should act.

2. Dismantling This Myth, Step By Step

The philosophical significance of the internal history of science is established by an argument like the four-step one presented below. After presenting this argument, I will systematically challenge each premise and short-circuit the inferential links between them.

(1) Most of the best cases of scientific reasoning have exhibited an independence, or "autonomy," from other sorts of delibera-

tions. This autonomy may be defined psychologically or institutionally, but in either case the general idea is that the deliberators do not let non-epistemic matters intervene in the epistemic assessment of a scientific claim.

(2) What makes these cases of scientific reasoning so good is their autonomy.

(3) The role of autonomy in increasing the likelihood that the most rational, objective, and/or valid scientific claims will be chosen has been sufficiently great to make autonomy worth recommending as a general methodological strategy.

(4) There is prima facie reason for believing that suboptimal cases of scientific reasoning—when less rational, less objective, or less valid claims were chosen—are traceable to the intervention of non-epistemic matters. From this tenet follows the division of cognitive labor canonized by Laudan (1977, ch. 7) as the *arationality assumption*: namely, that philosophers study scientific reasoning in its autonomous phases, whereas sociologists study it in its heteronomous ones.

Philosophers associated with "positivism" (Ayer 1959), "rationalism" (Hollis & Lukes 1982), and "scientific realism" (Leplin 1984a) normally differ on many things, but not on their allegiance to the above four tenets. In fact, virtually all these philosophers have written as if the tenets logically unfold in the order indicated, with (1) entailing (2), which then entails (3), which in turn entails (4). On the contrary, however, one could, as a matter of historical fact, accept (1) without accepting any of the other three tenets. Admittedly, this would not make for a very potent philosophy of science. Still, it is important to see that each successive inference could be blocked, if only to get an initial feel for the logical looseness of internalism and the problems that consequently lay ahead for it.

But let us begin on a positive note by granting both the intelligibility and truth of (1). The presence of autonomy in all the best cases of scientific reasoning must be supplemented by other inductive tests before the move from (1) to (2) can be licensed: Have an equally large number of suboptimal cases of scientific reasoning exhibited autonomy? And even in the best cases, might not different factors be responsible in each case for the reasoning's turning out to be as good as it was? After all, commonality of effect need not imply commonality of cause. If further empirical study produces affirmative answers to either of these questions, then the move from (1) to (2) is blocked, and the presence of autonomy will have been shown to be as epiphenomenal as the "fact" that all the best scientific reasoners have been males. However, assuming that these empirical hurdles are surmountable, there is a more trenchant conceptual barrier between (1) and (2).

The barrier involves getting a clear sense of what the *content* of a science might be. After all, if nothing else, "autonomy" implies the fixing of the will upon some object. In the case of autonomous science (or the autonomous scientist), the object seems to be knowledge, or epistemic content. But what exactly is that? The internalist tries to define content so as to separate it from anything obviously material which would make it susceptible to an externalist analysis. It is easy to think of this strategy as a safeguard against a form of weakness of the will that might befall the scientist who too closely associated the pursuit of knowledge with one of its material instantiations, such as the sheer production of texts formatted in a certain "disciplined" way, for which she was regularly rewarded. And so, true to its etymological roots, content is something "contained" equally in a set of texts and a set of practices, which is to say, it is not identical to either set of things (cf. Ong 1958, pp. 116–121). Indeed, the vast material difference between stable texts and fluid practices has perennially rendered the idea of content mysterious, manipulable, and ultimately suspect.

At this point, there are the makings of an unlikely triple alliance of doubters. An extensive discussion of the first camp—the literary and historical deconstructionists—may be found in Fuller (1988b, chs. 5–6). As for the other two, several philosophical students of cognitive science (Woodfield 1982, Pettit & MacDowell 1986) deny that there are any context-independent ways of specifying the content of beliefs, whereas social constructivists in science studies treat content as the rhetorical accomplishment by which two or more texts are mutually implicated in a discursive practice, especially in cases where the word of an authoritative figure alone manages to end any further debate on potential differences between the significance of these texts (i.e., what Latour 1987a calls "translating" or "black boxing"; Woolgar 1988b, ch. 3).

Moreover, the sophistication of these attempts to deconstruct the idea of content should not be underestimated. For example, Stephen Stich (1986) points out that the clarity of our intuitive judgment that two sentences have the same content (or are "synonymous") is no measure of the intuition's access to such content, since what we are inclined to count as being similar in content may be highly manipulable, even though the judgments themselves remain intuitively clear (cf. Quine 1960, ch. 2). In other words, merely feeling that one has grasped or expressed a certain content is not sufficient grounds for concluding that there is a content that one has grasped or expressed. This point is further developed in Chapters Three and Four, as we repeatedly encounter problems in trying to demarcate content from context in analyzing knowledge claims. Yet, for all their negativism toward content, neither the philosophers nor the sociologists on whom I rely should be taken as necessarily denying all forms of *realism*, an issue that is dealt with later in this chapter. At this point, it is enough to say that even if there is no determinate fact about what a scientist means (because there is no realm of meanings, propositions, or other content-filled things), there still may be a determinate fact as to whose authority may be invoked to justify the particular use of a scientific text in a particular situation.

The critique of content has serious implications for the conventional distinction between "internal" and "external" historiographies of science, since both sides normally agree that there is a content to science, but disagree as to whether that content may be studied independently of its ambient social history (cf. Roth 1987, chs. 6–8). True, internalists have tended to propose rather elusively Platonic criteria for this content, as in the case of the strong rational reconstructionists, Lakatos and Shapere. They have drawn the internal–external distinction so tightly that even the scientists' original formulations may be consigned to the non-cognitive external side, as language constitutes a noise-filled medium, useful for communicating with local audiences, but capable of obscuring the content that the scientists really meant to convey.

But the more interesting point to stress is that the externalists also assume that science has a content to be explained. Thus, if the three-pronged critique of content is correct, there is something misguided about the principles of "causality" and "symmetry" proposed by the Strong Programme in the Sociology of Knowledge (Bloor 1976), a group of self-styled sociologists of science at Edinburgh University who have been most effective in tackling the internalists over the past ten years. These principles direct the student of science to explain the content of science as she would any other cultural artifact. Unfortunately, cultural artifacts are more clearly bounded in space and time than content seems to be. In fact, the point that the Strong Programmers *should* be making is that "science," or better still, "knowledge," exists only as a manner of speaking, a way of analyzing social behavior—*all* social behavior—a certain way of collapsing contexts of utterance and abstracting from ordinary behavioral descriptions. Just as it is possible to talk about any social practice, from religious ceremonies to spectator sports to university research, in terms of its contribution to the flow of goods and services in the economy, it is likewise possible to talk about any of these practices in terms of its contribution to the production and distribution of knowledge. (A good place to start would be to chart the formation, maintenance, and change in agents' expectations of one another and their environments.) Admittedly, we are not nearly so practiced in, so to speak, the epistemification of the social world, for reasons that would make an interesting study in modern social taboos. But for that matter, the reception of, say, Gary Becker's (1964) *Human Capital*, indicates that we are not that much more comfortable with reducing the social world to economics. Still, the point remains: That banks and factories are explicitly defined as economic institutions does not entail that the economy transpires exclusively within their walls; neither should universities and other research environments be regarded as the exclusive domains in which knowledge is produced. In both cases, we just selectively focus our talk, and hence our attention, so as to ghettoize our understanding of society.

A good way of epitomizing the line of argument that I have been using to short-circuit the move from (1) to (2) is that uncritical appeals to the content of science are mired in the *textbook fallacy*. The particular textbook I have in mind is of the type used in most introductory sociology courses, which is organized around social institutions; in it the student finds chapters devoted to "the family," "the economy," "the educational system," and so forth. Because these texts do not normally theorize about the rationale behind their organization, the novice can be easily misled into believing that each chapter captures a discrete set of practices, associated with distinct places where they typically occur, which taken together make up the entire social order. Of course, however, if the textbook has any degree of sophistication, a student looking more closely will find considerable overlap among the practices described in the various chapters. For example, it is unlikely that a discussion of the family will be restricted to the formation and maintenance of *gemeinschaftlich* bonds. In addition, the reader is likely to find an analysis of the family as an economic unit, as a vehicle for transmitting political ideology, and the like. The point, then, is that, contrary to what one might naively think, the different chapters of the sociology textbook do not refer to types of social behavior but rather to types of analyses to which all social behavior may be submitted. And one such type of analysis, called "epistemic" or "cognitive," is most frequently used to describe cases in which what is normally called "science" is being done. But the empirical rarity of cognitive-talk about other forms of social behavior does not preclude its conceptual relevance to them, or them to it. In sum, then, the epistemic content upon which the autonomous scientist is fixed is a statistical mirage based on our ordinary unreflective patterns of speech.

But let us now ignore everything that I have said up to this point and imagine that the internalist makes it from (1) to (2). Needless to say, that still would not get her to (3)—and I do not even need to invoke the Humean taboos to make my case. That is, we need not prevent an "is" from yielding an "ought," or prohibit a "was" from becoming a "will be" (though the latter taboo may in fact turn out to be decisive, if the autonomous pursuit of science becomes too costly or risky). Rather, the problem is whether autonomy is something that can be *deliberately* achieved and, hence, something reasonably aimed for. It may be that although autonomy is indeed responsible for the best cases of scientific reasoning, autonomy can itself only arise as a by-product of the explicit pursuit of some other social practice. An example of this situation would be the (not entirely farfetched) discovery that all the relevant scientists had been engaged in long-term projects designed to promote their interests, during which what turned out to be their best moments of scientific reasoning occurred as tangents of the larger projects. In that case, the scientist's attempts to self-legislate autonomy by studiously avoiding non-epistemic matters may have some of the quality of deliberately trying to forget: both are psychologically impossible to adopt as an explicit epistemic policy (Williams 1973, ch. 9).

But to be true to the analogy, it should be observed that if the scientist's self-legislation of autonomy is anything like selective forgetting, then there may be indirect means by which he/she can achieve the desired effect (Elster 1979, part 2). For example, the scientist may conduct research in a "socially sterilized" environment, one from which all the reminders of non-epistemic matters have been eliminated. Plato's training ground for future philosopher-kings seems to have been an environment of this sort. More modern versions may include the laboratory far off the university campus whose researchers are so insulated from the rest of social life that they never even need to fill out a progress report to ensure continued funding. To design such a research environment would, of course, still be to pursue autonomy deliberately, but now via a circuitous route. And while this would establish the psychological possibility of adopting autonomy as an explicit epistemic policy, the desirability of such a policy—as stipulated in (3)—would have yet to be shown.

There are two reasons for thinking that the explicit pursuit of epistemic autonomy may be undesirable, even if possible. The first turns on the consequences of a likely state of affairs, namely, that the pursuit of autonomy is not entirely successful. Suppose, in other words, that the scientists cannot be completely insulated from the rest of society. Perhaps the society has allowed too many research teams to flourish, given the scarce resources available for their maintenance. Consequently, research teams must periodically compete for these resources, a situation that, in turn, makes the scientists susceptible to the interest groups with the greatest say over how the funds are distributed. Progress reports are thus tailored to appeal to these groups, even if it means distorting the significance of current research. Moreover, if the society's science policy is to allow scientists to pursue epistemic autonomy as much as possible, then these distortions could well go unnoticed, since the society would be averse to commissioning its own inspectors of the research sites, simply for fear of adding to the level of external interference.

Yet, as students of science policy are finding increasingly clear, once scientists are forced to justify their research in a public forum, the policy maker is more likely to make a sound funding judgment by holding the scientists accountable to on-site inspectors than by simply taking what the scientists say at face value (Campbell 1987a). This would suggest that if complete autonomy is unachievable, then the next best outcome may involve, not the minimal amount of social monitoring possible, but rather, a large amount of social monitoring of scientists' activities. This conclusion accords well with the principle of preference ordering that economists call the general theory of second best (Lipsey & Lancaster 1956; Goodin 1982, ch. 4). The idea is that if an outcome is preferred because it meets a set of conditions, and if one of these conditions cannot be met, then the next best outcome probably does not require meeting any of the other original conditions.

The pursuit of epistemic autonomy may be undesirable for a second reason, one that returns us to the issue of by-products—not autonomy itself as a byproduct, but the by-products of pursuing autonomy. One of the most telling arguments for showing that science in the United States today approximates a state of epistemic autonomy is that unless expressly required, scientists need not account for, nor even keep track of, how their research impinges on society at large. Left to their own devices in this way, scientists tend toward increasingly well-defined problems, whose hypotheses can be properly tested only after the construction of more elaborate technical apparatus. What the policy maker sees in this tendency is a future of more expensive and specialized research that draws both cognitive and material resources away from the problems of the public sphere, which are invariably amorphous and interdisciplinary in character. In short, one negative byproduct of epistemic autonomy may be the social marginalization of science, or what the likes of Bacon and Comte would bemoan as the severance of knowledge from power. Yet, to sever knowledge from power is not necessarily to destroy either one. Indeed, another negative by-product of science's epistemic autonomy may be the idle availability of those bigger, faster, and cannier things that scientists continually produce. As J. D. Bernal and other socially concerned scientists realized at the dawn of the Nuclear Age, the dark side of science's epistemic autonomy is the scientist's alienation from the uses to which her products are put (Bernal 1969, ch. 10). In other words, there is no neat way of protecting scientists from corrupt social influences that does not, at the same time, prevent the scientists from acting as salutary social influences (cf. Williams 1981, ch. 4, on "dirty hands").

But, finally, there is one overriding conceptual objection of going from (2) to (3). Just because someone intends to do something and succeeds at doing it, it does not necessarily follow that his/her intention played a significant causal role in bringing about his/her success. Georg von Wright (1971) vividly, albeit ungenerously, tagged this faulty inference, the *fallacy of animism*. It cuts against both internalist and externalist accounts of science that rely on psychologistic forms of causation. For example, externalists who show that scientists try and succeed at having their social interests satisfied fall by the wayside, as well as internalists who automatically trace the epistemic success of a research program to the autonomous, truth-regarding orientation of the successful scientists. Indeed, the fallacy even undermines something as basic as philosophical accounts of human communication that presuppose the successful conveyance of intentions (e.g., Grice 1957, Searle 1983). Derrida with a vengeance! The basic error involved here is to suppose that preparing one's mind in a certain way (cf. Descartes' "Rules for the Direction of the Mind") automatically improves the likelihood that events outside the mind will conform to the strictures of the mental preparation.

Once having made all these arguments against the move from (2) to (3), it might appear unnecessary to sever the inferential link between (3) and (4). But suppose, for the sake of argument, that we were able to countenance both the psychological possibility and desirability of autonomy as an epistemic policy. Would we then be permitted to move to (4)? Again the answer is no, at least not until the historian has deconstructed the carefully crafted Whig narratives that still dominate the historiography of science.

After all, a close scrutiny of the historical record may reveal that some of the best cases of scientific reasoning—autonomous ones at that—have been "reinterpreted" over the years because the conclusions that were originally reached did not contribute to the narrative flow of science. These reinterpretations aim, in part, to forestall the embarrassment that would come from our learning that the sort of autonomous reasoning that led, say, Lavoisier and Pasteur to their conclusions also guided Priestley and Pouchet to their quite opposite ones. The potential embarrassment here lies in the ironic implication that, in spite of their equally impeccable ratiocinations, what made the first pair "heroes" and the second pair "villains" (or better still, "victims") in the history of science were factors entirely beyond their control: namely, those who subsequently found use for their work. This would be the ultimate vindication of externalism! (The irony is compounded by noting that what Priestley and Pouchet could not control was who would end up writing the internal histories of science.) Consequently, there is a temptation for internalist historians to abide by the self-serving interpretive principle of presuming that the heroes are at least as autonomous as they seem (and hence there is no need to probe further), whereas the autonomy of the villains probably does not extend much below the surface.

But even granting the unlikely possibility that a properly deconstructed history of science would show that the "heroes" were indeed better reasoners, there are probably deeper considerations— pertaining to the nature of cognition itself—that would cast doubts on any easy diagnosis of suboptimal reasoning in terms of the interference of non-epistemic factors. Ironically, this objection comes from cognitive science, a disciplinary cluster that has so far pursued, from the standpoint of "methodological solipsism," a largely autonomous study of the mind. Paul Churchland (1979, ch. 18) has introduced the idea of an *epistemic engine*, a machine that learns how to learn over the course of successive encounters with the world. The machine's responses are informed entirely by a program of methodological rules and theoretical principles that will allow it to flourish in the world. But without any non-epistemic factors, such as passions and instincts, disturbing the epistemic engine's program, what exactly does the machine then need to learn from its worldly encounters? The panglossian answer given by Daniel Dennett (1987a, ch. 7) is that the machine still needs to learn when these rules and principles are appropriately applied. In other words, it is presumed that all errors made by the machine are due to the misuse of otherwise sound cognitive instruments. For example, once regarded as an epistemic engine, a scientist who failed to conduct falsifying experiments on a pet hypothesis would be understood, not as having succumbed to some extrascientific forces, but as having mistakenly applied another pattern of inference, say, modus ponens, which would be suitable on some other occasion. Thus, we have a scenario whereby epistemic factors alone, through their *mutual interference*, can be used to explain suboptimal reasoning. In fact, the scenario fits very well with the crypto-Hegelian historical sensibility of an internalist like Shapere, who sees the actual historical episodes as opportunities for the epistemic engine of science to become cognizant of the factors guiding its own course by internally differentiating the principles of reasoning governing its operation.

3. Gently Easing Ourselves Out of Internalism: The Case of Disciplines

Stephen Toulmin (1972) has pointed out that disciplines "interface" methodological and institutional matters and thereby define the border between internal and external history of science. He and fellow traveler Dudley Shapere (1984) have gone on to define the process of discipline formation as the routinized concentration of methodological resources on a specific domain of objects. So far so good. Yet, philosophers who have since taken up this approach have done so almost exclusively in order to illuminate the standard philosophical problem of identifying criteria for rational theory choice (e.g., Darden & Maull 1977; an exception is Bechtel 1986). The supposed virtue of studying disciplines in this context is that it brings the range of "good reasons" relevant for justifying a theory choice down to the level of the practicing scientist, which frees the philosopher from having to radically "idealize" the history of science in order to find examples of the philosopher's methodological maxims. Although it is difficult to show that many scientists try to falsify their own theories, it is easy to show that they try to argue in a manner that would be most persuasive to the target audience in their discipline. The reader may already begin to suspect that we are losing the institutional side of the interface that disciplines are supposed to define. Moreover, as we shall see, in looking at Shapere's work, the historical side is not doing too well, either.

Shapere grounds the enhanced epistemic power of modern science in its increasingly systematic and self-contained reason-giving practices. He is apparently oblivious to the fact that they are just one instance of the rationalized accounting procedures that have generally emerged in response to the complexity of modern society (Luhmann 1979). Yet, other institutions that have similar procedures—most notably the economy and the law—are not normally regarded as engaged in any knowledge gathering other than that of the people whose activities need to be accounted for. Thus, when Shapere (1987) says that the imperative to test theories is emblematic of the rationality of modern science, an institutionally oriented sociologist most naturally reads him as endorsing tight surveillance over the distribution of credit in the scientific community by requiring that scientists follow up their utterances with appropriate actions (cf. Collins 1975, chs. 6, 9).

In this context, the sociologist might continue, an important function of reason-giving is to lend objectivity, continuity, focus, hence legitimacy, to the practice in question, usually by eliminating any traces of contingency from accounts of the situations for which reasons need to be given (cf. Fuller 1988b, ch. 3). Here the translation of the research environment into the canons of scientific writing merits comparison with a judge's construction of court cases as "deductions" from existing laws (cf. Fuller 1988a). A telling trace of the contingency that scientific discourse tries to conceal arises in the definition of autonomy, a piece of doublethink that is needed to accommodate the idea that scientific research has become most self-directed during the period when it has been most subject to political and economic pressure (Turner 1986). That scientists typically do not make reference to "external" factors when offering reasons for their claims simply shows the extent to which they have gotten used to their self-imposed boundaries. (I discuss this point in more detail in the Coda.) Therefore, whereas Shapere likes to talk tough about "testing science on its own terms" or "making science live up to its own standards," on closer inspection, his words turn out to be euphemisms for taking canonical scientific accounting procedures at face value. As befits its suppressed contingency, Shapere's discussion of the history of science remains at an eerily Hegelian level of abstraction, in which the dates and the names function primarily as steps in a self-serving logical progression.

To be fair to Shapere, his focus on disciplines adds a measure of historical sensitivity to his internalist account. Indeed, he is notorious for arguing that there is no ahistorical core concept of science to be analyzed by philosophers (Shapere 1984, ch. 11). Yet, as we just saw in his privileging of science's reason-giving practices, Shapere does seem to think that the language of science represents the nature of science, ever-changing though it may be, in a fairly transparent fashion— certainly more transparently than the positivists had thought, who after all held that scientific claims had to be given some theory-neutral reformulation before their content could be determined. In fact, this point goes some way toward motivating Shapere's choice of disciplines (or "domains," as he often prefers to call them) as the proper focus of philosophical inquiry into science (Shapere 1984, chs. 13–14). Domains are the local ontologies and modes of epistemic access staked out by scientific disciplines. Although a domain ontology may undergo many changes, as the corresponding discipline changes, the contents of the domain at a given time can be determined by examining the theories that the local scientists deem to be the best.

Here Shapere is clearly trying to correct a tendency found not only among positivists but also among their natural foes, the scientific realists (about whom more later in this chapter). Both the positivist and the realist try to ground the legitimacy of their philosophies of science by invoking an imperialistic sense of "physicalism," where physicalism means, not so much a general commitment to science's being grounded in material reality, but more a specific interest in recasting all the sciences as branches of physics. However, this much may be said against Shapere: It is one thing to show that, as a matter of historical fact, disciplines locally determine the conduct of inquiry, often in disparate but productive ways, and quite another to show that there is no epistemically necessary or privileged way of conducting scientific inquiry. By defending the former rather than the latter, modal claim, Shapere fails to challenge the positivist and the realist on their own terms; for either sort of philosopher can simply conclude from Shapere's studies of domains that the drive to know is sufficiently robust that science can be satisfactorily, though not *optimally*, conducted even in the face of the human frailties revealed by the existence of disciplinary boundaries and the consequent fragmentation of inquiry. Next, Shapere would have to show that there is no optimal mode of inquiry, but only the array of modes that work satisfactorily in particular disciplines.

But in that case, is each of these satisfactory modes optimal in its respective discipline? If not, then why should scientists be local realists, rather than, say, social constructivists, about their domains? On the one hand, given his own historicist conception of science, Shapere should not want to say that the disciplines map isomorphically onto the structure of reality; yet, on the other, he certainly wants to avoid the constructivist move of reducing disciplines to plastic "opportunity structures" (Knorr-Cetina 1980). Is there a middle course to steer between this Scylla and Charybdis? It is significant that the historiography of science inspired by Shapere's work (e.g., Darden & Maull 1977) has tended toward showing how a theory or method developed in one discipline can satisfy another discipline's standing conceptual needs; at the same time, this historiography has neglected what practitioners of the provident discipline thought about both the epistemic status of the needy discipline and the appropriateness of the particular cross-disciplinary borrowing (for this neglected side, cf. Manier 1986). In other words, to get a feel for alternative historical interpretations within the actual history of science, and thereby stimulate a constructivist sensibility, an instructive move would be to write from the standpoint of the lending discipline, rather than from that of the borrower (i.e., the discipline that benefits), in an interdisciplinary transaction. Otherwise, it would seem then that Shapere's progeny are willing to err on the side of isomorphism over opportunism.

In contrast to Shapere, the social constructivists can make some headway against the positivist and the realist precisely because they *agree* with the philosophers that there is nothing epistemically special about the *canonical* division of cognitive labor into academic disciplines. What generally goes unnoticed in this debate is that the philosophers are especially *reflexive* on this point, since they hold that the contingent character of inquiry—the very need for disciplinary specialization—is itself the contingent product of the sheer magnitude of the scientific enterprise, which can be managed only by allowing networks of researchers a considerable degree of autonomy from other networks. Not surprisingly, these networks spawn relatively isolated bodies of discourses, which then become the bases for demarcating discrete domains of inquiry, à la Shapere. But who is to say that science must be as large and messy a social practice as it in fact is in order to achieve its end, namely, the representation of reality? This is the challenge that the philosophers pose with their socially streamlined physicalism. The social constructivists meet the challenge by showing that these unsavory features are integral to the nature of science: No science would be done, if knowledge claims were not the sort of things that had to be "staked out" by setting up disciplinary boundaries (cf. Bourdieu 1975, Gieryn 1983). *Where* the disciplinary boundaries are drawn may be conventional, but certainly not *that* they are drawn (cf. Fuller 1988b, ch. 8). Again, Shapere has little to contribute to this debate because he ends up conferring epistemic privilege on the canonical account of the development of science. He is thus that peculiar sort of apriorist (again, like Hegel) who sees history governed by the principle of sufficient reason.

In terms of his commitment to a social history of science, Toulmin fares not much better than Shapere. In fact, the only "institution" that Toulmin fully develops in his compendious *Human Understanding* is the forum, which is the "marketplace" (i.e., journals, conferences, and other media) where disciplinary practitioners try to instill in each other a demand for their ideas. Unlike, say, the more intellectually hermetic environment of Michael Polanyi's (1957) "Republic of Science," Toulmin's forum is definitely a disciplinary structure that is susceptible to external social forces. However, Toulmin treats these external forces as making a difference to the development of a discipline only insofar as they affect the course of debate inside the discipline's forum. Thus, Toulmin has much to say about the role of ideological interests in the arguments that scientists use to steer the course of future research, but nothing to say about the role that politicians, administrators, and other non-scientists (perhaps practitioners of other disciplines themselves) play in providing direction. Nor does Toulmin address what is perhaps the most interesting issue lurking here: namely, how one draws the line between a scientist arguing under the influence of ideological interests and a scientist simply arguing as a politician; or, for that matter, how one draws the line between a scientist arguing for a radical reorientation of research in his discipline and a scientist simply providing a radical critique of her discipline from the perspective of some other discipline.

Why does Toulmin presume that a sharp line can be drawn between the "insider" and "outsider" roles suggested in the preceding two contrasts? There is a sense in which this question is of some political importance within the science studies community. For although Toulmin's view is hardly dominant in contemporary philosophy of science, it has nevertheless received the sympathy of those philosophers, historians, and sociologists interested in an interdisciplinary study of science. Yet, Toulmin remains in the grip of a picture of disciplines that comes all too naturally to the internalist. His inspiration is most immediately traceable to that icon of liberal–positivist culture, the legal proceeding (cf. Hooker 1987, ch. 7): that is, a discipline is a set of procedures for introducing and contesting issues in a particular domain, one designed especially to neutralize potentially biasing factors in the deliberations. However, since disciplines supposedly have "lives of their own" that are devoted to representing some slice of reality, it may prove illuminating to fall back on a more organismic model. For lack of a better name, the picture I have in mind is one of *sociological phenomenalism*. On this view, a discipline is modeled on an autonomous ego for whom external forces exist only insofar as they can be internally represented as disciplinary phenomena. Thus, whereas government can influence science by having its opinion represented within the scientific community, government cannot simply circumvent scientific forums and intervene in its own inimitable fashion—for example, by diverting funds. At least, that part of the story would never be told by a sociological phenomenalist. This is why Toulmin and his associates have not found, say, the economics of research a relevant topic for discussion (cf. Rescher 1979). As far as they are concerned, if money talks, it is only with the aid of an epistemic translator.

The hidden appeal of sociological phenomenalism in contemporary debates in the philosophy of science should not be underestimated, for it suggests a general strategy for enhancing the plausibility of an internal historiography of science, namely, to represent in suitably internalist terms—that is, in terms of the relevant disciplinary forum—many of the phenomena that would normally be taken as externally caused. (I imagine that this also explains the current allure of *autopoietic* accounts of science in German spheres of influence: cf. Maturana & Varela 1980.) We can regard this strategy from the standpoint of either the internalist historian or the scientific agent herself. After acknowledging the persistent economic, political, and miscellaneous social pressures that are exerted on scientific judgment, the internalist historian still wonders, especially in cases where the scientist seems to have made "the right choice," whether these pressures did little more than preempt the normal operation of scientific methodology. The historian's implicit causal claim would then be that, notwithstanding their ubiquity, external factors are not necessary to the conduct of inquiry.

For her own part, the scientist under study would regard the matter more opportunistically. Under the sway of external factors, the scientist would undertake research commitments that she perceives as likely to maximize her interests, though she would publicly justify that decision in terms appropriate to the internal operation of her discipline. Thus, what the historian takes to be the end of her inquiry, namely, the autonomous development of disciplines, becomes, in the hands of the scientist, the means for bringing about some desired outcome. In short, the strategy of disciplinary boundary maintenance (Gieryn 1983; Fuller 1988b, ch. 7) that underlies sociological phenomenalism plays to both *prospective* (i.e., the scientist's) and *retrospective* (i.e., the historian's) senses of rationality. Putting the matter somewhat crudely but vividly: Newton realized that mathematical demonstrations were needed to get physicists to accept the sort of principles that vindicated his Hermeticist interests; the internalist historian, however, realizes that Newton's demonstrations would have stood on their own epistemic merit, even without his having been motivated by Hermeticism. As we shall see in Chapter Three, Laudan's recent work is sociological phenomenalism with a vengeance.

4. If Internalism Is Such a Myth, Then Why Don't the Sociologists Have the Upper Hand?

The "sociological phenomenalism" of Toulmin and Shapere clearly demonstrates both the lure and the limitations of philosophical attempts to understand science as a social, specifically disciplined, search for knowledge. The source of this clarity lies in sociological phenomenalism's nearness to the mythic roots of internalist approaches to science. Thus, as I shall now argue, the internalist–externalist dispute that has engaged philosophers and sociologists of science in recent years is best understood as replaying the classical solutions to the *mind–body problem*. My opening move toward making this point is to consider the seemingly self-contradictory sound of "social epistemology," which suggests both an empirical sociological enterprise and a rationalist philosophy of science. The label is thus a veritable microcosm of the impasse that currently paralyzes theory development in science studies. However, once it is appreciated how an enterprise can be both "social" and "epistemological" at the same time, new strategies for theory development can be launched. As it turns out, bridging the social and the epistemological will be no easier—and no more difficult—than bridging the physical and the mental.

Suppose a moderate member of the Strong Programme in the Sociology of Knowledge (cf. Bloor 1976) were asked to define the terms of the debate between herself and a rationalist philosopher of science. How would she respond? I hope that the reader will find nothing especially controversial about the following fabricated remarks, since they will serve to highlight the extent to which certain pernicious assumptions have been uncritically accepted by all sides:

> Whereas the philosophers think that sociology can only explain the emergence and maintenance of non-rational scientific beliefs, we sociologists hold that the resources of our discipline can be used equally well to explain the formation of rational scientific beliefs. Admittedly, the philosophers are correct in observing that scientific texts are most naturally read as concerning the proposal and appraisal of rival accounts of some extra-social reality. However, sociologists have found that such a reading is far too superficial, neglecting as it does the background social interests that determine what is proposed and how it is appraised, but that, because of disciplinary writing conventions, never find their way to the surface of the scientific text.

The first, rather positive point to make about this passage is that it nicely captures the "naturalness" of philosophical readings of the history of science, which force the sociologist to show that there is more to the scientific text than meets the eye (McMullin 1984 exemplifies this rhetoric). Thus, sociologists engage in a "hermeneutic of suspicion" (Ricoeur 1970), warning us not to be fooled by the seemingly self-contained intertextuality of scientific discourse, and laboratory ethnographers present themselves as going "behind closed doors" and "reading between the lines." Philosophers have tended to take this interpretive stance as indicating that the sociologists realize that they must shoulder the burden of proof in establishing the relevance of their analyses to an understanding of science. Needless to say, the philosophers are being somewhat self-serving, since another, and probably more plausible, way of justifying the sociologist's suspicious stance is as a methodological reminder not to fall into an elementary scholastic trap of confusing what is first in "the order of knowing" (i.e., what immediately strikes us as demanding explanation) with what is first in "the order of being" (i.e., what provides such an explanation). The relative autonomy of scientific reason-giving practices from the rest of the social order is certainly a striking phenomenon, but as with other striking phenomena, there is no reason to think that it can be adequately explained by remaining at the level of epistemic first impressions.

In fact, strange as it may seem, the sociologist's hermeneutic of suspicion is really just the humanistic variant on what would normally pass for "realism" in the philosophy of the natural sciences (cf. Fuller 1988a). Both the sociologist and the scientific realist are devoted to penetrating the appearances, typically for penetration's own sake, and certainly with little regard for whether the established beliefs of scientists or society at large remain intact. In contrast, philosophers who continue to study science in purely "internalist" terms are like instrumentalists in the natural sciences, whose ultimate aim is to "save the phenomena" as simply as possible, given the uses to which they would put such phenomena, and regardless of whether those uses are ever likely to reveal the real mechanisms that make the phenomena possible. Of course, the "use" to which philosophers put their accounts of the internal phenomena of science is in developing theories of rationality, which, once legitimized by science, can then be extended to evaluate, and perhaps even regulate, society at large (cf. Rouse 1987, chs. 6–7). In short, the moderate sociologist in the passage may have needlessly betrayed a defensive posture toward the philosophers, which suggests that she has failed to see the irony of her situation. For as I have been arguing, the picture of science as a self-directed, self-contained activity is the product of philosophers' adopting a self-interested, instrumentalist stance toward science: They stand to gain the most professionally, as keepers of rationality, if science is presented as "naturally" being nothing more than the sum of its textual appearances.

Pursuing this line of reasoning a little further, let us now call into question the very "naturalness" of the philosopher's reading of the history of science. How often is it said—in support of the philosopher's position—that internalist history of science simply requires that you know what the scientists are talking about (i.e., the "content" of the particular science), whereas the history preferred by sociologists requires the introduction of factors that are extrinsic to what the scientists are saying? The force of this argument, which again puts the sociologist on the defensive, turns on a characteristically philosophical way of thinking about cognitive content as something detachable from social context, much in the way mental properties have been seen as something detachable from brain states. And in both cases, the detachment is specious. The problem, however, is that we are more used to recognizing the speciousness of the mind–brain split than of the content–context split. Just as in the heyday of dualism, when mind and brain were taken to be distinct (albeit interactive) entities, philosophers and sociologists of science are still prone to think of the cognitive and the social as mutually exclusive (though perhaps complementary) sets of "factors." (To his great credit, Bruno Latour [e.g., 1987a, 1987b] has made significant strides toward eradicating this picture.)

To hear some philosophers tell it, you would think that one of the things that can be determined from closely observing scientists in action is whether a social or a cognitive factor is operating at a given moment. For example, is the scientist appealing to rational argument, or is she merely buckling under social pressure? These two states are presumed to have quite different empirical indicators, just as consciousness is supposed to be the mark of the mental, whereas the absence of consciousness allegedly points to automatic physical response. The truth of the matter, though, is that "the social" and "the cognitive" are not separate parts of the scientific enterprise; rather, they are two relatively autonomous discourses that are available for analyzing *any* part of science.

What I am claiming here—and again by analogy from the mind–brain case—is that any instance of scientific behavior can be described either "cognitively" or "socially," depending on the discursive context in which the analyst's description is embedded and, consequently, the sorts of generalizations that she is interested in eliciting. Something that would "naturally" be called a case of testing a hypothesis is at the same time analyzable as the mobilization of certain kinds of political and economic resources (i.e., the capital and labor that need to be in place to operationalize the test). Likewise, something that would "naturally" appear to be an instance of class interests biasing scientific judgment may equally be analyzed as a cognitively sound strategy (i.e., "satisficing," to use Herbert Simon's term) for buying some time until more conclusive testing clarifies the relative merits of the rival theories (cf. Giere 1988, ch. 6).

Contrary to some Marxist and Durkheimian analyses of cognitive relations as the abstract "reflection" of social relations (e.g., Bloor 1982), I do not maintain that every instance of hypothesis testing can be mapped onto the same sort of social practice. Instead, my point is that there is a social practice (not necessarily the same one) to which each instance of hypothesis testing corresponds. I am, thus, arguing for what philosophers of mind call a "token–token," rather than a "type–type," identity theory of the social and the cognitive for the domain of science (cf. Churchland 1984, ch. 2). What this means is that I envisage social-talk and cognitive-talk as *orthogonal*, rather than *parallel*, classifications of the same phenomena, that is, scientific behavior. The two cross-classifying discourses operate as schemes for subsuming different features of scientific behavior under internally related, but mutually exclusive, sets of laws (cf. Fodor 1981, ch. 5). In the case of cognitive-talk, these laws probably include the sort of methodological principles that philosophers of science typically study, along with the more discipline-bound ways in which scientists rationalize their own activities to one another in print. In the case of social-talk, the relevant laws range over the entire gamut of psychological, sociological, political, and economic variables that are normally used to explain any other sort of social behavior. I shall have more to say about this when I discuss the various recent attempts to "naturalize" philosophy of science.

Two questions arise about my proposal. First, how exactly does one establish these token–token mappings between social- and cognitive-talk? To say that both the social and the cognitive articulate the common phenomena of "scientific behavior" is to say nothing yet about the level of abstraction, or "molarity," at which the two articulations occur. This is an especially pressing issue in the case of cognitive-talk, where philosophers have been traditionally vague about whether individuals or groups (and of what size and shape?) are supposed to be the vehicles for methodological principles (cf. Campbell 1979). Although full discussion of this topic must be delayed until the next chapter, we may now say, in anticipation, that if the proper vehicles are not chosen, it then becomes all too easy to show that science is a methodologically unprincipled activity.

The second question concerns whether social-talk and cognitive-talk are truly on an equal epistemological footing. Readers familiar with recent turns in the debates over the mind–body problem (cf. Churchland 1984) know that token–token identity is usually treated as a conceptual way station on the road to a more satisfying solution (Fodor 1981 is an exception). In fact, there has been a tendency to see brain-talk as having the epistemic upper hand over mind-talk, in that brain-talk's categories are the ones more likely to be integrated into a plausible overall world-talk (i.e., physicalism). I think that something similar can be said about the epistemic advantages of social-talk over cognitive-talk with regard to some overarching theory of human behavior. If so, it would follow that the relative autonomy of cognitive-talk from social-talk is a form of false consciousness that systematically distracts its speakers (i.e., rationalist philosophers of science, as well as many scientists) from situating science in the greater scheme of human things. The "naturalness" of cognitive talk is, therefore, just as much the product of artifice as introspection's special access to the mind is now claimed to be (Lyons 1986). Indeed, had the people responsible for establishing the history of science as an academic discipline in the early twentieth century been, say, literary historians or ethnographers, rather than physical scientists (e.g., Pierre Duhem, Ernst Mach) and Platonist philosophers (e.g., Emile Meyerson, Alexandre Koyre), the most "natural" reading of scientific texts would pertain to rhetorically important stylistic nuances and implicit speech acts, which are certainly no farther metaphorically from the textual surface than propositional content.

5. Still, the Internalists Do Not Have a Lock on the Concept of Rationality

We need to prod our imaginations into thinking of alternative conceptions of scientific rationality. A good stimulant is Harry Redner's (1986, Appendix) potted history of the concept of reason. Redner argues that there have been three ideal types of reason in Western culture, which have been combined in various proportions over the last 2,500 years. As it turns out, there is also a trend in this history, namely, toward reason becoming increasingly "externalized" and (so Redner seems to think) less integral to the human condition. Adapting Redner's categories, it is easy to see the sort of movement he has in mind:

(a) *Reason is inherent in the world as it ordinarily is*: To understand
the nature of something is to know its purpose in the overall
design of the world. Thus, the natural simply is the normative.
This is the Greek conception of *telos*, which in humans takes
the form of *logos*, which allows us our measure of self-
directedness. This conception survives intact pre-eminently in
Natural Law jurisprudence as practiced by, say, the Roman
Catholic Church. It instills in inquirers the sense that
everything in the world is good, if understood in its own terms
and not forced to perform a function for which it was not
designed.

(b) *Reason is inherent in the world but must first be released*: To
understand the nature of something is to know how it is under
ideal conditions, which do not ordinarily obtain, but which
can be released from ordinary conditions by the pursuit of
some disciplined activity, or "method." This is the Enlightenment conception of *raison* and *Vernunft* as the force that frees
humans from the bondage of ignorance of both themselves and
the world. Notice that reason in this sense crucially vacillates
between being what is released by the method [which brings it
closer to (a)] and being the method itself [which brings it
closer to (c)]. This conception survives as the scientific
attitude and its attendant "modernist" sensibility: While the
world is not necessarily as we would like it, how we would like
the world to be is within our reach, once we uncover its
underlying nature.

(c) *Reason is not inherent in the world but must be imposed from the
outside*: To understand the nature of something is to impose
some conceptual order on its activity that would otherwise be
lacking. Thus, both the natural and the human worlds are, in
an important sense, resistant to reason. This is the twentieth
century conception of *rationalization* as a principle that
contains and pushes back the natural onset of mental and
physical disorder. It operates most notably as *sublimation*,
whereby the prima facie experience of irrationality is taken to
be merely a transient phenomenon that is best understood in
terms of its contribution to a long-term, large-scale plan. And
although this plan is normally of one's own design, it could
also be the design of someone else (cf. Freud 1961).

If nothing else, this scheme illustrates the slippery character of
rationality, as (a), (b), and (c) easily blend into one another. Indeed,
only a tortured point of theodicy prevents (c) from becoming

self-mystified, and thereby reverting to (a): the difference between (a)'s panglossian sense that everything *has* a purpose, whether or not we realize it, and (c)'s more self-deceptive sense that everything *needs* to be given a purpose on account of the world's inherent unintelligibility (cf. Elster 1984b, part 4). But more to the point at hand, Redner's scheme provides a convenient means of locating where the social epistemologist's conception of scientific rationality stands vis-à-vis that of the philosopher under the spell of the internal history of science. Whereas the internalist straddles (a) and (b), I straddle (b) and (c). In what follows, the internalist's straddle is called "inertialist"; my own is called "impressed."

Rationality may be regarded either as something *released* or as something *imposed*. That is, rationality may be seen as releasing a deep order discernible from the apparent chaos of human existence or as imposing an artificial order on our chaotic nature. The former is clearly the sense of rationality that an internalist historian of science presupposes, whereas the latter is the sense that an externalist like Bruno Latour endorses under the rubric of rationality. Let us conjure up some vivid images of what is at stake here. The idea of "releasing" someone's rationality is reflected in the administering of intelligence tests to solitary individuals placed in a quiet environment, removed from the company of other people and without such cognitive aids as books and calculators. By contrast, the idea that rationality is "imposed" makes sense upon noticing that even a scientist is reduced to babbling incoherence when taken out of a habitat of other scientists, books, blackboards, and instruments, all of which normally serve to constrain the range of responses that she makes.

If we were back in the eighteenth century and interested in designing a "mechanics of the mind," we would cast the contrast as being between *inertial* and *impressed* forces of reason, and would appeal to Descartes and Hume as our respective authorities. On the inertial view, humans are naturally predisposed to rational thought and will therefore attend to the dictates of reason unless they are subject to external interference. The role of method, then, is to remove any potentially extraneous thoughts from the mind so that reason may freely flourish. On the impressed view, however, humans have only one mental predisposition, namely, to follow whatever attracts their attention at a given moment. They do not so much strive for reason as they are swayed by passions. In that case, rationality involves subjecting one's naturally wayward thoughts to explicit rules of order.

The epistemological distinction made a political difference in the aftermath of the French Revolution of 1789. As the economist Thomas Sowell (1987) has argued, the distinction reflects two visions of human nature, the *unconstrained* (cf. inertial) and the *constrained* (cf. impressed). Thus, whereas the unconstrained Tom Paine fretted about oppression inhibiting the workings of Natural Law in the human spirit, the constrained Edmund Burke worried about chaos ensuing from the renunciation of civil law. An important advantage of reformulating the inertial–impressed distinction in these terms is that no expertise in late eighteenth century rhetoric is needed to know that whether Paine's or Burke's vision prevailed, both were calling for political action that would certainly seem constraining to the ideologically naive—indeed, perhaps even to the not so naive.

Finally, within science itself, inertial and impressed forces of reason made the most difference in defining the two main research traditions of nineteenth century social science. On the inertial side was the Anglo-German paradigm, whence came the sciences of neoclassical economics and experimental psychology. These were generally *naturwissenschaftlich* inquiries that located rationality in any healthy individual human being, who was portrayed as a coherent utility maximizer. On the impressed side was the Franco-German paradigm, whence came the sciences of sociology and political science. These were generally *geisteswissenschaftlich* inquiries that located rationality in the community as whole, with each individual being the locus for a cluster of conflicting role expectations (Manicas 1986).

But are the policy implications of the two views as distinct as these conceptual differences suggest? On the one hand, the inertialist recommends constructing the "initial conditions," so to speak, for enabling the emergence of our natural rationality. These may include specific mental training or simply being placed in an ideal environment, such as Plato's Academy. Once these conditions are in place, the reasoner is characterized as exercising a certain freedom of thought, or "autonomy." On the other hand, the impressivist argues for more directly bringing about a certain desirable pattern of behavior deemed "rational." If that involves coercing and manipulating people and discourses in various ways, so be it; for without the constant application of force, rationality would once again return to disorder. If nothing else, a clear difference of explanatory strategy is implied here.

The inertialist vindicates her view by noting that no set of initial conditions can fully determine the reasoner's response. Consequently, if the reasoner responds uniformly enough to an appropriate variety of these conditions (imagine the different forms that the intellectual life can take), then it would seem that our inherent rationality is being tapped. By contrast, the impressivist takes the alleged underdetermination of the reasoner's response by initial conditions to be an artifact of our not fully understanding all the conditions that have determined the response. Moreover, the alleged uniformity noted by the inertialist is explained away either as the product of blurring subtle differences in the behavioral response or as having been triggered by subtly similar features in the initial conditions. At this point, inertialism should start to resemble the view championed by a nativist like Noam Chomsky, whereas impressivism should start to look like B. F. Skinner's behaviorist brand of empiricism. In the former case, the reasoner comes across as able to generate an amazingly rich response to an impoverished environment, whereas the latter case portrays the reasoner as having gotten used to managing, in a largely subconscious but expedient manner, an environment fraught with more stimuli than can possibly be attended to.

Following the lead of Egon Brunswik (1952) and other psychologists concerned with the "ecological validity" of accounts of the organism as a creature of its environment, it might be argued that the relative persuasiveness of inertialist and impressivist explanatory strategies will hinge on the richness of the languages available to describe rational behavior. Let us first take a brief look at how this point makes a difference in assessing research in contemporary psychology. Here laboratory-based experimentalists display a subtle nativist bias by presuming without question the physicist's latest understanding of matter in motion, quantum mechanics, and then asking how is it possible for us to a perceive a world so stable that we not only survive but also find it commonsensical. Given this disparity between a meaning-impoverished external world and a meaning-enriched mind, it should come as no surprise that the research agenda of most experimental psychology is guided by questions of what enables us to transform the atomic swirl around us into meaningful objects. Whatever particular answer is settled on, it will certainly impute a highly ramified structure to the mind (Devine & Ostrom, 1988). The psychologist James Gibson (1979) has tried to demystify this research agenda by calling for the psychologist to do her own physics, one tailored to the levels at which the organism under study can perceive physical reality, for example, an "ecologically valid optics." Baron (1988) has since laid the groundwork for an ecologically valid theory of social knowing that reinterprets many of the attribution errors that subjects make in traditional social psychology experiments.

In a similar vein, the internal history of science promotes an inertialist conception of rationality by recounting the recurrence of methodological truths amid a volatile and unsystematic realm of "social factors," as if the historian can simply make do with the methodologist's (i.e., a philosopher of science's) sense of methodology—especially the abstractability of method from concrete research contexts. This, in turn, encourages the historian's research question to be framed around the persistence of method against the noise of history. Not surprisingly, the range of admissible answers will invoke some transcontextual epistemic feature of these methods, such as truth-enhancement. Demystifying the internalist problematic requires debunking the inertialist model of rationality in two steps. It involves, on the one hand, showing that the alleged persistence of method is an artifact of restricting historical description to a relatively abstract level, namely, to the utterances (and sometimes paraphrases) of key scientists and philosophers. A finer-grained analysis of these utterances, taking into account changes in context as well as ordinary semantic drift, would reveal a variability comparable to the rest of social reality. On the other hand, the pockets of invariance in methodological practice that remain after this analysis is performed are to be explained by features common to, not transcending, the social situations under study.

Different images of knowledge follow from the inertialist and impressivist viewpoints. To illustrate this difference, let us consider one area where these images have had major import, namely, in the debates over the nature of the market's "regulation" in capitalism. By the late nineteenth century, it had already become apparent that the expanding market economies of Europe and the United States were not self-regulating mechanisms on the model of mature organisms and closed physical systems (Lowe 1965, ch. 3). Though there were cycles of booms and busts, there were no discernible equilibrium points for such key indicators as rates of unemployment, inflation, and income growth, points toward which these economies naturally gravitated. Given the political strife that this volatility could surely breed, the question arose as to whether markets, after having reached a certain rate of growth and level of complexity, needed some "external," specifically governmental, regulation to maintain their stability. How one answered this question turned very much on the interpretation that one gave of the volatility of the marketplace: Was it a matter of misinformed economic agents unwittingly destabilizing the system, or rather a matter of well-informed agents continually shifting the points of equilibrium in the system? In contrasting the Keynesian and Austrian answers to this question, we will get a very clear sense of the dynamics of Redner's ideal types: On the Keynesian side, there is the inherent instability of a (b)-type rationality, whereas on the Austrian side, there is the tendency of a (c)-type rationality to revert to an (a)-type.

Historically, the inertial view has been represented in Keynesian and other planned approaches to the economy (Lowe 1965, ch. 9).

These approaches tend to suppose that there is a set of facts about the overall state of the economy at any given time (including the conditions under which it functions optimally), of which the economic agents have varying degrees of knowledge, by virtue of their particular positions in the market. Moreover, there is no reason to think that in the normal course of their exchanges, the agents will improve their knowledge of these facts, since day-to-day success in the market rests more directly on knowing what others think is going to happen in the economy—however misinformed most people may be—than on knowing what the objective indicators predict. And although the objective indicators may ultimately prove the majority wrong, it is doubtful that most economic agents would remain in the market long enough to realize how wrong they were, unless they could correctly anticipate each other's misinformed views in the short term. Consequently, the systematic, or "macro," character of the economy is reserved to the economist as a special realm of knowledge, to which the economic agents themselves have only limited access and to which they only unwittingly contribute through the product of their exchanges. It is a realm populated by objective indicators, such as rates of inflation and unemployment, divested of psychological trappings and distilled into real currency flow. Armed with the economist's knowledge, the state planner can compensate for the errors and excesses of the economic agents by constraining their exchanges in various ways to produce an optimally functioning market.

In the Keynesian example, the key features of the inertialist image of knowledge emerge. First, the "inertial mass" whose natural motions the economist wants to capture is that of the entire economy, and the information that she provides the state planner consists of policies to compensate for the faulty psychological tendencies of the economic agents whose interactions constitute the market. In addition, the line drawn between the reality known to the economist and the appearances in which the agents dwell is, so to speak, metaphysically "hard," much like Plato's "divided line" in Book VI of the *Republic*. In other words, the Keynesian economist does not expect that any long-term improvement in the sort of knowledge had by economic agents will ever converge on the sort of (superior) knowledge had by the economist. This drives home the familiar point that an economics degree does not necessarily improve the chances of success in the market (and vice versa, it might be added). In fact, if the economist's degree is to have any efficacy at all, she needs to occupy a social position *away from* the market, informed by non-market interests and attendant to data to which the economic agents do not have access—again much like the philosopher-king in the *Republic*.

By contrast, the Austrian neoclassical school of economics, most comprehensively represented by Friedrich von Hayek (1985), captures the impressivist image of knowledge. On the Austrian view, central economic planning is doomed to failure because it tries to replace the unique knowledge that economic agents have of their own talents, interests, and local market situation with theoretical abstractions, which, in the final analysis, simply summarize concrete economic activity drained of the very concreteness that makes it all work. If the economic agents do not operate from the categories that ground the economist's knowledge, then that should not be interpreted as a cognitive shortcoming but merely as a sign that the agents have had no need to frame their activities in terms of those categories. Whereas the Keynesian would remark that all that this shows is the self-perpetuating character of ignorance, an impressivist like von Hayek would argue that the collective activity of the agents implicitly—as an instance of the invisible hand—determines the level and sort of knowledge that is appropriate to flourishing in the marketplace, to which the agents then naturally adapt. Therefore, if it is fair to characterize the Keynesian as presupposing the "a priori ideality" of the economy, to which real markets are made to conform through central planning, then the Austrian economist presupposes the market's "a posteriori ideality," in other words, whatever the economy does is what the economy ought to do—or, in baldly Hegelian terms, "the real is rational, and the rational is real."

From this brief consideration of the foundations of economic knowledge, we can already begin to see that the ease with which the difference between "releasing" and "imposing" rationality can be destabilized reflects a potentially deeper instability about what is "real" versus "apparent," "objective" versus "relative." It is with complete deconstruction in mind that I now turn to the remaining sections of this chapter.

6. Nor on the Concept of Reality, Where Things Are a Complete Mess

When philosophers nowadays debate "the status of science," they are no longer questioning the legitimacy of science—especially natural science—as a way of pursuing knowledge. That point has been tacitly conceded for more than a hundred years. Rather, philosophers are typically interested in the *extent* and *source* of science's legitimacy. One of the most curious features of the ensuing debates has been that the plausibility of a given position depends very much on how the status of science is formulated as a point of contention. Quite similar positions bear quite different burdens of proof, depending on the context of the debate. This can be easily illustrated by considering the two main formulations of the science debates. The first, favored by philosophers who are normally concerned with the analysis of scientific methodology, may be rendered as follows:

Are there legitimate grounds for criticism in science aside from those having to do with judgments of empirical adequacy?

Realists say yes, while *antirealists* say no. That is, the realist argues that there may still be grounds—internal to the aims of science—for challenging a scientific theory, even if the theory accounts for all the available data. The antirealist denies this point, though she may hold a variety of views on the nature and frequency with which scientific theories actually do "save the phenomena." As this debate currently stands, the realists have the upper hand—at least the antirealists behave as if they had to bear the burden of proof. The antirealist typically defends her position by arguing that the theoretical terms that figure so prominently in science are really nothing more than heuristics for pursuing additional data, despite the realist's claim that theory captures what lies beyond the readily observable. Thus, the antirealist's fate rests on the viability of Ockham's Razor as a regulative principle of scientific pursuit, which then permits the realist to be portrayed as an ontological spendthrift.

However, in the second version of the science debates, the state of play appears somewhat different. Here is its formulation, which is preferred by "post-analytic" and "hermeneutical" philosophers who study science as an integral part of contemporary culture:

Are there legitimate grounds for criticism in science aside from those having to do with judgments of expert authority?

Objectivists say yes, while *relativists* say no. That is, the objectivist argues that there may still be grounds—internal to the aims of science—for challenging a scientific theory, even if the theory has received the unanimous support of the scientific community. The relativist denies this point, though she may take a variety of positions on the nature and frequency with which scientists actually reach consensus in their deliberations. In this debate, the relativists have the upper hand—at least insofar as "objectivism" is a term generally applied to others and rarely to oneself. Consequently, objectivists try to improve their position by arguing that relativists paint a superficial picture of science, one that simply takes for granted that a difference in language reflects a difference in concepts (and perhaps even objects), which, in turn, neglects the long-term continuities in our cognitive enterprises, especially the persistent interest in representing reality.

The reader will no doubt notice a parallel between, on the one hand, realism and objectivism and, on the other, antirealism and relativism. Why does this parallel cut in favor of one side in one debate and the other side in the other? At the very least, it suggests that the two debates are rarely conducted together. And, indeed, the reader is hard-pressed to find mention of "objectivism" and "relativism" in the writings of such recent luminaries in the realism–antirealism debates as Bas van Fraassen, Larry Laudan, Richard Boyd, and Ian Hacking, all of whom are concerned with the historical trajectory of the natural sciences (Leplin 1984a). Likewise, talk of "realism" or "antirealism" is not often heard from Richard Bernstein (1983), Alasdair MacIntyre (1984), or the various students of Popper and Wittgenstein who have wrestled with the objectivism–relativism issue in debating the foundations of the social sciences (Wilson 1970). For example, while ever the objectivist, Popper's attitude toward "realism" has ranged from contempt, when regarded as essentialism in the social sciences (Popper 1957), to wholehearted endorsement, when treated as three semi-autonomous realms of being (Popper 1984). By contrast, in his later writings, Wittgenstein (1958) emphasized the local basis of epistemic standards and is thus normally treated as a relativist. However, Michael Dummett (1977) gave Wittgenstein an Oxford twist, pumping up his philosophical anthropology into a full-blown antirealist metaphysics, which has been especially influential in the philosophy of mathematics (e.g., Wright 1980). Finally, Richard Rorty (1979, 1982) and Hilary Putnam (1978, 1987) are interesting examples of philosophers who have gradually moved from the realism–antirealism debate to the objectivism–relativism one. In both cases, terms from the former debate drop out almost entirely when the philosopher is engaging in the latter. Moreover, this tendency goes beyond Rorty's and Putnam's mercurial philosophical personas. It reflects an admittedly vague intuition that however closely related they may be, "realism" and "objectivism" cannot be used interchangeably, nor can "antirealism" and "relativism." But before probing the differences, we must get the similarities straight. What is the common ground, on the one hand, between realism and objectivism and, on the other, between antirealism and relativism?

Realism and objectivism share the view that there is more to science (more precisely, scientific validity) than meets the eye or the mind of the scientist (or the scientific community). Antirealism and relativism concur in denying this point. In the past hundred years, philosophers—most notably Charles Sanders Peirce (1964) and Karl Popper (1972)—have tried to arrive at the best of both positions in the science debates by constructing the realist and objectivist side as the limiting case of the antirealist and relativist side. Thus, these philosophers have defended the claim that science appears to be an especially "realistic" or "objective" form of knowledge because scientists aspire to the standards of empirical adequacy and consensus formation that would characterize the scientific community at the logically ultimate stage of inquiry. On this view, realism can be, so to speak, "simulated" by showing that when scientists are at their best, they aim to account not merely for what is currently observable but also for what may be observed only after many more years of systematic study, namely, the "underlying causal mechanisms" that are the stock-in-trade of realist analyses of science. Similarly, objectivism can be simulated by showing that scientists aim to convince not merely their immediate colleagues but also imaginary inquirers detached from contemporary social pressures and devoted to the pure pursuit of knowledge.

7. The End of Realism, or Deconstructing Everything In and Out of Sight

Now that we have established that the sides of the two science debates share some common ground, it should be possible to show that realism's strength in one debate and objectivism's weakness in the other have resulted from historical accident and not from something inherent in the two positions. A good way of making this point is by resorting to a dialectical strategy familiar to devotees of deconstruction (Culler 1983). Deconstructionists are best seen as radical positivists interested in revealing the meaninglessness of philosophical controversies. But whereas positivists (e.g., Ayer 1971) typically proceed by showing that no empirical difference is made by taking either side of a dispute, deconstructionists try to persuade the reader that no conceptual difference is made by taking sides. Thus, according to the positivist, the science debates cannot be resolved because their partisans are unable to specify any difference that a given resolution would make to the actual conduct of science (e.g., would it change the pattern of theory choices, revealing some apparently bad choices to have been good ones, and vice versa?). In contrast, the deconstructionist would trace the irresolubility of these debates to each side essentially presupposing the truth of the other. In that case, it is not a matter of, say, choosing realism or antirealism, but rather of choosing realism-and-antirealism or opting out of the debate entirely.

I shall sketch how this strategy works for the four positions in the two science debates and then consider the actual tactics more closely. To deconstruct the antirealist, the realist needs to show that empirical adequacy or verifiability is not a self-sufficient concept but one that involves implicit reference to a reality beyond the phenomena. For her part, the antirealist can deconstruct the realist by establishing that realism is itself motivated by some notion of empirical adequacy. Similarly, the objectivist deconstructs the relativist by pointing out that the relativist is committed to an objective notion of social, if not physical, reality, which renders intelligible the idea that there is a "fact of the matter" as to who has cognitive authority in a given society. And finally, the relativist can return the favor by pointing out that the very idea of objective reality only makes sense relative to beings whose cognitive grasp is sometimes exceeded.

The realist makes her move against the antirealist by first capitalizing on the latter's self-effacing epistemic posture. Antirealists typically admit that the verifiability conditions of some theoretical entity can never be completely determined; for this would involve specifying how the entity would appear under all the empirical circumstances in which it could appear. The easiest way to imagine all these possible verifications is, following Francis Bacon and Ernst Mach, as the table of values that would satisfy the variables in a physical equation such as Newton's Force Law or Snell's Refraction Law (Mach 1943). Admittedly, physics has left the overwhelming majority of these cases untested, yet it is presumed that these cases would lend support to these laws, were they tested. What licenses such a presumption on the part of the antirealist, asks the realist, if not a tacit commitment to a world that continues to exist even when unrevealed?

The realist can press her point further by noting that what in fact passes as conditions of empirical adequacy in both normal and revolutionary science is even sketchier than the table of values example would suggest. Scientists deal in knowledge claims, not by exchanging recipes for arriving at unique experimental outcomes, but by setting mutually acceptable constraints within which permissible outcomes must fall. Moreover, the constraints are fairly loose—usually just consistency with relevant principles and data—and are themselves subject to revision as scientists monitor their efforts (Nickles 1980b). The cognitive situation of the empirical scientist can thus be likened to a couple of people chasing down a book in a library, when one person has caught a glimpse of the book and gives the other person vague directions as to its location. However, in the course of their search, the two people first run across another book that does the job of the original book, and decide to settle on it as their source of information. The realist would note here that had the search been driven primarily by the *empirical* properties that the first person had glimpsed of the original book, the two people would not have then had adequate grounds for ending their search with the second book. However, it is clear from the story's conclusion that the search was in fact driven by certain other properties that were only contingently related to the empirical properties glimpsed initially. These other properties, pertaining to the book's content, are like the "reals" that the realist takes to be the aim of scientific inquiry. Not surprisingly, in this context, realists are prone to claim that the first person did not merely register the surface features of a particular book, but rather "intuited" the *kind* of book of which both books in the story were instances (Putnam 1983).

But let us now turn the tables and see how the antirealist would deconstruct the realist. We shall end by retelling the tale of the two books. At the start, though, the antirealist grants the realist's point that conditions of empirical adequacy in science are much looser than the antirealist would like. However, she would be quick to add that these conditions, however inadequate, are still decisive for making sense of the realist's project. After all, the postulation of so-called inferred (as opposed to directly observable) entities in science, such as underlying causal mechanisms, largely takes the form of a promissory note redeemable upon bearing sensory witness to these entities. Rom Harre (1986), for one, has based an attenuated version of scientific realism on precisely this tendency which he has detected in the history of science: that is, the tendency toward rendering the unobservable observable. If, after enough time, the inferred entities have not been made manifest, the entities have typically been discarded. The antirealist concludes from this that the realist is able to think that her position underwrites scientific progress only by taking too short-term a view of the history of science. The realist focuses on the inadequate empirical conditions that existed at the times when scientists have introduced inferred entities, while neglecting the fact that the staying power of these entities has depended on their being issued substantial empirical warrants in the long term.

In this regard, the tale of the two books looks persuasive only because it is a case in which the inquirers were issued such warrants upon finding the second book. At the same time, however, the tale draws our attention away from the many cases in which the original glimpses were not borne out in the subsequent search, as well as the subtler cases in which the inquirers settle for something essentially different from what they originally sought (Fuller 1987a). If we keep these alternative endings to the story in mind, so argues the antirealist, we then realize that the realist's sense of "intuition" is little more than the (fallible) ability to predict the eventual observability of an entity that is currently only inferred.

Turning now to our other debate, the objectivist begins her deconstruction of the relativist by distinguishing relativism from solipsism, the view that the jurisdiction of cognitive authority extends no further than oneself. The objectivist notes that relativism historically emerged as a methodological framework for determining the standard of cognitive authority in various societies (Rosenthal 1984, ch. 8). The strategy basically involved identifying the group to whose expertise the natives deferred on a given range of issues. Now, the constitution of this group and the nature of its expertise clearly varied from society to society. Yet, for each society, the relativist could, at least in principle, identify the relevant groups and the extent of their cognitive authority.

One implication of this view, observes the objectivist, is that the experts acquire many of the "objective" properties formerly ascribed to "nature" or "the world as it is in itself." Thus, it is still possible for rank-and-file members of a society to have false beliefs about some subject matter, if these beliefs do not correspond to the received word of the relevant experts. Note the truth idiom of "correspondence" here. In fact, on such esoteric matters as the nature of physical reality, most members of our own society—even if we assume epistemological relativism—turn out to hold false beliefs, since they are not knowledgeable of the latest physical theories. If pressed, the relativist would redescribe the historically unprecedented cognitive authority conferred on the natural sciences in terms of many societies, which differ substantially on other authorities, concurring in their deference to the same, albeit geographically dispersed, community of scientists. Admittedly, the relativist would explain such convergence by citing local factors that made the difference in the case of each society. But this does not undermine the objectivist's general point, which is to show that the relativist presupposes a notion of objective reality—one that is located in particular societies rather than nature at large (cf. Roth 1987, chs. 6–8).

And for the last round of our deconstructive debate, consider the relativist's revenge on the objectivist. Gottlob Frege unwittingly launched the opening salvo when he set the precedent for arguing that only bits of language—we would now say sentences—could have the property of being true or false, referring or not referring, in virtue of standing in an appropriate relation to some extra-linguistic reality (Dummett 1977, ch. 8). In so defining truth from the language end of the language–reality relation, Frege tipped the dialectical balance in favor of the relativist. Presumably, all natural languages are epistemic peers, in that they are of equal descriptive adequacy to the needs of their respective communities. After all, even the slightest inadequacy can be remedied by adding a new word or usage. And, on anyone's notion of descriptive adequacy, most of what is said in these languages must be true. Yet these languages are not readily intertranslatable. The most natural conclusion would seem to be that each language articulates a different world. As evidence, consider that the bilingual speaker finds it easier to switch between the two languages of his competence than to translate from one to the other. Indeed, this was one reason that Kuhn gave for advancing the incommensurability thesis and proposing that if we envisage scientific paradigms as akin to natural languages, then researchers who switch paradigms are moving between different worlds (Kuhn 1970b, p. 267). Given the intuitive implausibility of this conclusion (at least to objectivists), efforts have been made throughout this century to construct ideal formal languages and translation schemes to overcome the specter of relativism. However successful one judges these efforts to have been, the relativist's point remains that the objectivist has clearly assumed the burden of proof in attempting to show that bits of language are true or false of one and the same extra-linguistic reality.

But the relativist can deconstruct the objectivist from a more classical metaphysical standpoint, one that has come back into vogue with the emergence of Science & Technology Studies (Latour 1987b, but also Campbell 1987c). The idea of a reality that exists independently of our conceptions arises from the fact that certain things seem to escape our cognitive grasp (i.e., our beliefs and our desires) and resist our concrete attempts at transforming them. Here we are calling forth the various day-to-day disappointments, ranging from mispredicting the weather to failing to get the umbrella open. Indeed, the very word "reality" evolved from the Latin for being resistant to change. If we had perfect cognitive and practical control over everything that interested us, there probably would not be a need for distinguishing an objective reality. However, the exact extent of our imperfect control, and hence the line dividing objective from subjective reality, has shifted throughout the centuries. This is a point that figures prominently in Husserl's (1970) later work and Gaston Bachelard's (1984) philosophy of science, in both cases traced to Galileo's attempts to separate out "primary" qualities intrinsic to an object from the "secondary" qualities imposed upon the object by the subject (Koyre 1978). The mark of the primary quality was its invariance under the changing conditions of perception (e.g., the light shining on the object). But with the instrumental enhancement of perception, and the increased ability to manipulate more properties of the object, the character of primary qualities changed as well. Although these tendencies do not destroy the concept of objective reality, they highlight the extent to which the concept has been implicitly defined relative to the definer's cognitive and practical capacities—or rather, liabilities.

8. But What's Left of Scientific Rationality? Only Your Management Scientist Knows For Sure

So far the resolution of the realism–antirealism and the objectivism–relativism debates does not look promising. It is clear that realism and objectivism, on the one hand, and antirealism and relativism, on the other, seem to interpret the history of science in distinctly different ways. One benefit of this observation might be that we could collapse the two debates into one with only the rhetorical trappings lost. However, we earlier saw that the two sides of each debate, once dialectically deployed, also tend to collapse into each other. Relativism and objectivism presuppose each other, as do realism and antirealism. Although it would be fair to say that many interesting issues are raised in the course of arguing any of the four positions, which explains their continued centrality to the philosophy of science, the fact remains that none of these issues brings the two debates any closer to a resolution. Perhaps what we need to do, then, is to find a point *outside* the terms of these debates that will nevertheless help us sort out the confusions *inside* the debates.

Our first clue is to be found in the furor that followed Popper's (1957) use of "historicism" in *The Poverty of Historicism*, a polemical critique of positivist and Marxist attempts to arrive at universal laws of historical progress. Critics of the book noted that Popper's alternative methodology for the social sciences—which would have the inquirer explain human behavior in terms of locally occurring factors to which the agent could, at least in principle, have conscious access—bore a strong resemblance to what had previously been called "historicism," the species of relativism associated with Wilhelm Dilthey and modern hermeneutics (Hoy 1978). This brings out an important point about the significance that relativists attach to history. For what matters to them is not that history permits the identification of repeatable processes, but that it permits the specification of unique events. In the words of Max Weber and other Neo-Kantian theorists who, at the turn of the century, launched relativism on its current wave of epistemic respectability, history provides *idiographic*, event-oriented knowledge, whereas sociology provides *nomothetic*, process-oriented knowledge (Harris 1968, chs. 9–11). But whereas the Neo-Kantians generally regarded the two sorts of knowledge as complementary, twentieth century social scientists have often drawn correlative distinctions with the express purpose of discrediting the nomothetic side, thereby leaving relativism unchallenged and untempered. Most influential in this regard has probably been Claude Levi-Strauss's (1964, ch. 1) distinction between the *synchronic* knowledge afforded by the methods of contemporary ethnographic fieldwork and the *diachronic* pseudo-knowledge that the early anthropologists thought could be divined from ordering societies on a unilinear evolutionary scale.

The perceived strengths and weaknesses of relativism are directly traceable to its association with an idiographic (or synchronic) sense of history. Relativism works best as a strategy for studying epistemic communities whose pattern of change, if discernible at all, is indifferent to the order of events in time. In such communities, there is no evidence of the sort of epistemic process that would be implied by the net "growth" or "loss" of knowledge over an extended period. And precisely because the order of events in these communities does not proceed in any epistemically significant direction, the events can be safely studied as a mere succession of occurrences. In that case, whatever changes occur from event to event can be adequately explained in terms of specific local factors, without invoking laws alleged to apply to all instances of such change. From this brief description, it should be clear that the relativist's stronghold is the sort of community of which field studies are typically done: so-called static or primitive societies, where changes occur haphazardly and tend to cancel each other out in the long run.

That having been said, it must be immediately observed that over the past fifteen years the relativist framework has been imported into the study of Western scientific communities. This project, which goes under the rubric of *social constructivism*, is nowadays the type of sociology of science most familiar to philosophers and other practitioners of the humanities (Gilbert & Mulkay 1984, Gergen 1985, Woolgar 1988b). Since the pattern of epistemic change in scientific communities is not normally seen as indifferent to time order, critics in both philosophy and sociology have, not surprisingly, found the social constructivist accounts radically incomplete. Indeed, the earliest—and most notorious—accounts in this vein typically omitted any discussion of what happens to a piece of knowledge once it leaves the perceptual and conceptual horizons of a well-defined group of scientists, whose movements have themselves been monitored only within the confines of a laboratory. To most students of scientific knowledge, this is to leave off where the most interesting part of the story begins: namely, the part that describes how that piece of knowledge is subsequently used, both to extend the knowledge base and to enhance society's technical capability.

Both philosophers and sociologists of science have ways of characterizing the realms into which the social constructivist dares and dares not venture. As the philosopher sees it, the social constructivist avoids the context of *justification*, the public forum that has traditionally been the philosopher's turf. Rather, the constructivist confines her activities to the more private context of *discovery*, on which philosophers have only sporadically dared to tread (Nickles 1980a). Whereas philosophers speak of the context of discovery as fraught with the arbitrariness and idiosyncracies of "individual psychology," they take the context of justification to embody a "logic" of some sort that permits a knowledge claim to be evaluated in a standardized fashion, thereby rendering the particular place and time of the evaluator, at least in principle, irrelevant. Indeed, time enters the picture only in the processual, evolutionary sense normally disparaged by the relativist, in that philosophers tend to believe that a by-product of the fact that science has been done for many centuries is that scientists—and especially their philosophical onlookers—have been increasing their understanding of the comparative worth of rival justificatory strategies. This, in turn, enables later scientists to give their inquiries a clearer focus. Philosophers aid this process by articulating the ideas implicit in the scientists' greater understanding. We have already picked up some of this sentiment in our discussion of Lakatos's and Shapere's rational reconstruction of the history of science.

By contrast, sociologists mark the distinction in contexts less self-consciously, mainly by switching the social science from which they draw their accounts of scientific knowledge production. Whereas discovery is indeed discussed in the *anthropological* terms favored by the social constructivist, justification tends to be presented in terms borrowed from *political economy*. For example, in his recent *tour de force*, the French sociologist Bruno Latour (1987a) starts by explicating the face-to-face negotiation of a knowledge claim in one laboratory and then proceeds to chart the claim's cycle of "epistemic credibility" as it circulates through the society-at-large. Latour portrays scientists as always trying to construct and amass "immutable mobiles," which function in knowledge production much as money does in the production of economic goods. Immutable mobiles are knowledge claims that have gained sufficient independence from their original contexts of discovery that scientists with quite different interests (both epistemic and practical) can enhance their standing by invoking them. These mobiles are most evident in the literature reviews at the start of most scientific journal articles, in which the scientist reconstructs a research lineage that justifies her undertaking the problem to which the rest of the article is devoted. Latour makes much of the fact that these reviews rarely go into the detail of earlier work, nor do they sort the work into categories familiar to philosophers of science, such as "theories," "methods," and "data." He concludes that what matters is only that *many* items of such work are cited, and that they are items often cited by others in their work. This reinforces the economistic thrust of Latour's treatment of justification, one that values quantity of a highly desired good (i.e., a highly cited article) over the intrinsic quality of the good (i.e., whether what is claimed in the article is true). Moreover, it marks a significant break with how philosophers have conceived of scientific justification—a break that perhaps provides the key for getting beyond our two debates.

Philosophers typically treat the justification of a knowledge claim as a matter of seeing whether the claim has conformed to some generally applicable methodological norms, whereas a sociologist like Latour tends to regard justification as a post facto honor conferred on a knowledge claim that has managed *by some means or other* to survive the vicissitudes of the intellectual marketplace. (Michael Polanyi [1957] had held a similar view, except that he demoted honor conferred on the justified knowledge claim to one of *popularity*.) As we have just seen, both approaches run counter to the relativist sensibility of the social constructivist, in that they remove justification from the conceptual and perceptual horizons of the scientists who originally constructed the knowledge claim. But in addition, Latour's sociological approach challenges the standard philosophical assumption that justification, as opposed to discovery, is epistemically indifferent to the place and time of its occurrence. Since philosophers think that scientists are primarily interested in such intrinsic features of a knowledge claim as its truth, these features are portrayed, not surprisingly, as accessible, regardless of where and when the justification is attempted. By contrast, the Latourian sociologist would go so far as to argue that *timing and placing*—that is, the reception of a given knowledge claim by several key but unrelated audiences—may tell most of the story of whether and how the claim succeeds.

On the sociological view of justification, philosophers are captive to wishful thinking when they suppose that a claim is widely accepted for reasons that are themselves widely accepted—a wish, after all, that is behind the very idea of a universal logic of justification. On the contrary, as historical research has repeatedly shown, mutually incompatible methodological reasons—let alone pragmatic ones—have been invoked by members of a scientific audience for accepting a knowledge claim (cf. Fuller 1988b, ch. 9). Indeed, the main philosophical arguments for the realist and antirealist approaches to science are themselves based on alternative justifications that key scientists have given for accepting what is purported to be the *same* theory, with Newtonian mechanics the most frequently invoked case. To get a sense of the incompatible justifications offered by these two approaches, consider that realists saw Newtonian mechanics as *achieving* the aims traditionally pursued by metaphysics, whereas antirealists saw it as *replacing* those aims with more pragmatically oriented ones (Laudan 1984, ch. 3). If philosophy of science is defined as the search for a logic of justification, then the discipline continues to make sense in the face of the historical record, only if one holds that some of the scientists had *better reasons* than others for jointly accepting what turned out to be the better theory. And, as Peirce and Popper would have it, to determine these better reasons, scientists would need to leave (at least hypothetically) their local research contexts and defend their theory choices to a full forum of their potential peers. In this scientific version of an "ideal speech situation" (Thompson 1984, ch. 9), purely local interests would no longer be relevant, thereby freeing scientists of any need to press self-interested claims beyond their merit, which, in turn, paves the way for universal agreement. The underlying assumption here is noteworthy: the probability of consensus formation is increased as the communication channels between research teams is opened.

However, the more we flesh out the search for justification in sociological terms (and it is not clear how else the concept could be fleshed out), then the idea that justification embodies anything as dignified as a "logic" becomes problematic. For if the art of building coalitions is no different in science than in any other social practice, then what is likely to get disparate parties to accept a given knowledge claim as justified is more a matter of appealing to the "lowest common denominator" of the target parties than a matter of appealing to standards that transcend all parties in their universal applicability. In terms that a linguist could appreciate, the justificatory language is bound to look like a pidgin (i.e., a trade language), not a Chomskyan grammar (cf. Fuller 1987b). Brian Baigrie (in private communication) has pointed out an interesting example of this phenomenon. Why accept Newtonian mechanics? Some Newtonians originally appealed to its mathematical innovativeness, whereas others appealed to its association with experiment. The appeal to experiment won the day, largely because the relevant Newtonian experiments could be reproduced with relative ease, whereas few could follow Newton's mathematical proofs. Now, in the twentieth century, the justificatory tables have been turned, since advanced mathematics has been successfully integrated into the physics curriculum, but the complexity of contemporary physical experiments makes them comprehensible only to physicists who have actually participated in them. Consequently, mathematics is the lingua franca of today's debates in physics. Of course, some philosophers (and physicists, for that matter) may want to infer that this is because mathematics plays a universal role in scientific justification. But the coincidence with the fact that mathematics has become comprehensible to the physics community at large is just too close to ignore.

For their part, sociologists take the quest for a logic of justification to be a philosophical fairy tale that can be subverted with a few tips from Kuhn (cf. Barnes 1982). Far from being detrimental, the relative isolation of research communities from one another is actually instrumental in bringing about whatever sort of consensus scientists manage to reach on a theory. Why? As the relevant forum becomes more local and restricted, the scientist faces the less onerous task of justifying her theoretical choices to an audience—her immediate colleagues—whose ethos she understands and probably shares. The choices made through these deliberations are then incorporated into the discipline's body of knowledge (i.e., journals, books), which is subsequently made available to other research teams who may find the first team's epistemic contributions interesting for quite different methodological and pragmatic reasons. However, these differences in the reasons for appropriating a theory need never be made evident to the research teams concerned. In this way, scientists are able to circumvent an obstacle that has perennially prevented philosophers from engaging in collective projects, namely, the philosophers' felt need to agree on what counts as good reasons for adopting a theory prior to agreeing on whether a particular theory is good (Schlick 1964).

To appreciate the incapacitating character of this piece of philosophical metatheorizing, one need only turn to a paradigm in crisis, when scientists are faced with diagnosing and treating an anomalous finding. At that point, they quickly fall into a "philosophical" mode of arguing, whereby previously hidden differences in their justificatory strategies rise to the surface of discussion, often with the result of throwing into doubt whether divergent research teams are really talking about the same theory. This is another basis of Kuhn's (1970a, chs. 7–8) notorious incommensurability thesis. The philosopher still in search of a global logic of justification will probably find something strange about the sociologist's scaled-down alternative. In particular, she may detect a curious metatheory of rationality at work in the sociologist's call for local justificatory canons. The metatheory may be formulated as a three-step recipe:

(1) Start with the fact that scientists do not disagree about every feature of a knowledge claim and assume that the potential for disagreement increases as the claim is subject to finer-grained levels of semantic analysis. (Thus, there is no disagreement that $F = ma$ is Newton's Second Law of Motion, that F stands for force, etc. The potential for disagreement arises when one tries to give the meanings of "force," etc., and increases when one tries to specify the roles that $F = ma$ may play in scientific argument [cf. Buchdahl 1951].)

(2) Restrict discussion of the knowledge claim to contexts in which the potential for disagreement either is avoided altogether or is explicitly resolvable. (Thus, scientific journal articles avoid the discursive definitions of concepts as much as possible, unless a method is given—via experiment or calculation—for operationalizing the discursive definition. Another way of avoiding "deep" disagreements is by *showing* that one's own knowledge claim follows from or is presupposed by the knowledge claims of others, without *saying* how those claims must be interpreted in order for the connections to be made. Thomas Nickles [1985] has developed this point as the "generativist" account of scientific justification.)

(3) In the event of an anomalous finding, when scientists are forced to discuss changes that need to be made to the structure of knowledge in their discipline, make sure that regardless of how radical the changes turn out to be, they will tend to minimize the appearance of anomalies in the future. (Thus, a key reason why scientists are willing to engage in philosophical arguments about interpreting the anomaly one way rather than another is the prospect of avoiding similar arguments in the future. This explains why, in the cases of "scientific revolutions," scientists have moved to adopt a new paradigm instead of patching up the old one that promises only more anomalies in the long run.)

The strangeness of this metatheory—especially as a metatheory of *scientific* rationality—lies in its attitude toward disagreement, which is more one of *avoidance and containment* than the usual philosophical attitude of *endless encounter*. However, this failure to meet philosophers' expectations does not itself reflect badly on the metatheory, which finds ample support from the annals of science. An especially vivid case of this metatheory in action was the agreement by the founders of the Royal Society of Great Britain, in exchange for being granted a charter from the king, not to experiment in metaphysical, religious, and political matters. Once the effects of scientific discussions were so contained and their potential for being entangled in public policy debates avoided, the members of the Royal Society took further steps to define the parameters of scientific debate by introducing the concept of the journal, and especially that of a board of gatekeepers at the editorial helm (Bazerman 1987, ch. 5). Indeed, the ideology of science as an "autonomous" activity in the pure pursuit of knowledge arises just when scientists begin to exercise such self-restraint (Schaefer 1984).

In short, it would be difficult, given the metatheory sketched above and its substantial historical precedent, to see how Popper, Peirce, and other defenders of global approaches to justification could argue that the scientist's rationality is simply the philosopher's critical attitude continued by technically enhanced means. Rather, the sociologist has presented us with an account of *bounded rationality* whose exemplars are such beleaguered beings as the bureaucrat, the investment broker, the military strategist (March 1978). The bounded rational agent is a gamesman of sorts, but in a sense that is propaedeutic to the kind of "language games" that has come to signal Wittgensteinian approaches to knowledge production (e.g., Winch 1958). Whereas the follower of Wittgenstein (1958) envisages the rationality of the knowledge producer to lie in her ability to conform to the standing rules of her tribe, the knowledge producer conjured up by the sociologist's metatheory is rational in virtue of her ability to bound the field of epistemic play, so to speak, by setting down rules where none had previously existed (Bourdieu 1975; cf. Brown 1978).

Being a bounded rational agent involves nothing so heroic as what Popper or Peirce would suggest, namely, that the scientist gamble on a hypothesis that if true would eliminate many competitors but that more likely will, under severe test, eliminate only itself. Rather, granting that she is likely to end up having been mistaken, the scientist ought to protect herself from the worst possible outcome. But such a strategy does not require that the scientist select only "conservative" hypotheses, in the sense of ones that stray little from what is already accepted. On the contrary, as (3) makes clear, a scientific revolution may be, in certain cases, the best strategy for achieving the appropriate sense of "conservation," namely, of the scientific enterprise itself, which, in practical terms, means that only a limited number of scientists stand to lose a limited amount of epistemic credibility by the end of a major scientific controversy. This ensures that enough scientists will remain to engage in similar (or preferably, less divisive) controversies in the future. In short, where disagreement cannot be avoided, the effects of being on the losing side must be contained. With these issues in mind, Donald Campbell (1987b) has recently revived Francis Bacon's point that the rapid growth of knowledge that has taken place since the seventeenth century may be seen as following from a revolution in the adjudication of knowledge claims that was calculated to end the need for any such revolutions in the future. By the requirement that all knowledge claims survive the test of experiment, most of what had previously passed for knowledge was called into question; but since then most of what has passed the test of experiment has been retained as "phenomena" for which any scientific theory in that field has been held accountable (cf. Hacking 1983, ch. 13).

9. Finale: Some New Things for Philosophers to Worry About

However, the philosopher may remain unsatisfied with the sociologist's metatheory for deeper reasons. The philosopher could point to the sociologist's studied refusal to keep separate what the philosopher sees as "merely pragmatic" and "truly epistemic" aspects of scientific reasoning. The sociologist seems to be exclusively concerned with manufacturing an environment in which consensus is likely to emerge, regardless of whether the consensus is epistemically justified. Moreover, given my earlier remarks about the local character of justification, it is not even clear that the sociologist adheres to the spirit of "giving evidentiary reasons for one's beliefs," the core idea of epistemic justification. I shall argue, however, that what appears, from the philosopher's standpoint, to be a mixture of pragmatic and epistemic aspects of justification is, from the sociologist's standpoint, a distinction between justification as a *short run* and as a *long run* activity.

To start with the short run, if by "reasons" we mean the propositions that the scientist sincerely takes to support her conclusion, then these will address the interests of the scientist's research team. For example, the scientist might argue that since the team is better equipped to carry on research assuming the truth, rather than the falsity, of a particular hypothesis, and since any more attempts at its falsification would consume a vast amount of time and money, it is in the team's best interest to halt further testing and publish its results as conclusive. In what sense, if any, are the reasons offered in this argument "non-evidentiary?" At first glance, it would seem that the scientist is trying to convince her colleagues to forgo the strictures of method; but if so, how much more evidence (and of what sort) should she have encouraged them to collect? The philosopher would no doubt respond by advising that enough evidence be collected until only one reasonable hypothesis is left standing. She would then show how to perform crucial experiments, via a strategy of "strong inference," so that one or more rival hypotheses can be definitively eliminated with each round of testing (Tweney et al. 1981, part 3). Such is the way of epistemic justification. Unfortunately, if this advice were taken seriously, the team's research program would probably be defunct by the time the relevant tests were completed—that is, if they were ever completed at all. For whereas philosophers are quite good at recommending which hypothesis to endorse, they have little of use to say about the all-important issue of *when* an endorsement should be made. This point underscores a fact generally missed by philosophers overly concerned about the purity of epistemic justification: Whatever may be the scientist's primary short-term epistemic interest, it is *not* the attainment of truth-at-all-cost.

What, then, is the scientist's short-term epistemic interest? The sociologist's metatheory suggests that the scientist is interested in demonstrating the viability of her team's research program, ideally so that others are persuaded to make use of what the team publishes (cf. the concept of *interessement* in Callon et al. [1986]). Of course, in the *long run*, truth and the other classical epistemic virtues may be instrumental in the maintenance of the research program. But the sociologist's point is that these virtues cannot be successfully pursued as explicit short-term goals, precisely because they have no well-defined short-term outcomes. The philosopher herself admits as much in her reluctance to see the research team's decision to publish as anything but (epistemically, if not pragmatically) arbitrary. Still, the philosopher is faced with the question: If the research team is ultimately trying to pursue truth, or maximize coherence with other bodies of knowledge, or attain some other epistemically virtuous state, what exactly should its short-term expectations be? As Nickles (1987) points out, philosophers have only been recently canvassing for answers.

But the sociologist, drawing on her account of scientists as bounded rational agents, has an answer, one that retains the idea that the reasons scientists offer for endorsing a hypothesis are evidentiary. The twist is that the sociologist would have the evidence do double duty for the scientist: The scientist uses evidence not only to judge which hypothesis is most likely true, but also to judge whether more evidence is likely to alter that judgment. In other words, evidence functions as both a natural and a social indicator—equally a measure of the likelihood that the data are representative of the object under study and of the likelihood that the scientific community will sustain the scientist's interpretation in the long run.

However, contrary to what might be the philosopher's thinking on this matter, the sociologist does not see these two sorts of judgment as separable at the time a given hypothesis is evaluated. Rather, they provide alternative ways of reading the significance of the research team's decision after it has already taken effect. The more the team's credibility has been subsequently enhanced, the more likely the team's decision will be seen as having been informed by an astute reading of natural indicators; whereas the more the team's credibility has diminished, the more likely the decision will be seen as having failed to gauge the drift of scientific opinion (Latour 1987b). Advanced students of the role of causes and reasons in human action will recognize this phenomenon from arguments for the possibility of "bringing about the past" by some action in the present or future (Horwich 1986, ch. 6). At the level of psychology, this phenomenon is a case of what Jon Elster (1986, pp. 20–21) has called the "Everyday Calvinism" of the scientist's historical sensibility.

The practice of interpreting the utterances of an alien culture offers a telling analogy for the role of historical distance in clarifying the distinction between "merely pragmatic" and "truly epistemic" justification. This should come as no surprise, since the conditions for interpretive practice are generally implicated in any theory of rationality (Davidson 1983). The analogy is brought out in the following example. A sentence originally uttered in Attic Greek using the word *chrysos* simultaneously defined an area of semantic space (i.e., the conceptual relations of *chrysos* to other substances recognized by the Greeks) and picked out a set of objects in real space (i.e., things containing gold). But as later generations of interpreters came to realize that the Greek conception of gold was flawed, the difference between the semantic and the real space mapped by Attic Greek grew clearer, which is to say, the Greeks were gradually seen as having a conceptual framework separate from the interpreter's, a framework whose semantic space was an unreliable guide to real space. When the Greeks said things about gold that were still regarded as true, it was attributed to their having perceived features of real space, and hence their words having "referred" (cf. "truly epistemic"); whereas, when the Greeks said false things, it was attributed to the limitations of their conceptual framework, and hence their words having functioned in a purely "performative" (cf. "merely pragmatic") capacity (cf. Fuller 1988b, chs. 4–5).

From the standpoint of the debates over realism and objectivism, the interesting feature of our conclusion is that it breaks with a pair of assumptions common to the two sides of the two science debates in which we have been entangled throughout second half of this chapter. On the one hand, it breaks with the assumption, common to relativists and antirealists, that scientists define the criteria by which *their own* activities are evaluated. Rarely, if ever, do the partisans of these two positions countenance a situation in which an epistemic standard is *both* endorsed by a specific group of inquirers *and* applied primarily to inquirers outside that group without the consent of the outgroup's members. Yet, this is precisely the sort of control that later inquirers exercise when they reconstruct the original research team's decision-making process. On the other hand, the sociologist's conclusion also clearly breaks with the objectivist and realist assumption that "how things are" is independent of, or indifferent to, the constructions of scientists—but again not for reasons that would sit well with a relativist or an antirealist. For the above example also suggests that there are varying degrees of independence that the truth may have of a given scientist's construction; roughly, ontological independence seems to increase with spatiotemporal distance. Thus, the truth about the research team's activities (i.e., its ultimate contribution to the overall course of inquiry) is less dependent on what the team itself thought than on what the later inquirers turned out to have thought.

Reposing the Naturalistic Question:
What Is Knowledge?

1. Naturalism as a Threat to Rationality: The Case of Laudan

If Larry Laudan (1987) had his way, philosophers would salvage the concept of scientific progress at the expense of the concept of scientific rationality. You might say that in this way the proponents of naturalistic and normative approaches to the philosophy of science would reach a state of peaceful coexistence: the normative elements would retreat to their home turf of present-day cognitive interests, whereas the naturalistic elements would be allowed free passage in providing historical episodes that, in retrospect, can be shown to have (or have not) promoted those interests. The rhetoric of reconciliation aside, Laudan is really restricting the normative dimension of the philosophy of science to the only area where it would seem to be ineliminable: namely, where it concerns our own interests in wanting to understand the history of science. These interests constitute an implicit theory of progress, which once articulated, can explain why some past theory choices appear historically significant, others less so, and still others ironically so.

In his earlier work, Laudan (1977, ch. 5) wanted to capture the "pre-analytic intuitions" of rank-and-file scientists who came to realize that a particular research tradition is worth pursuing over its competitors. While several philosophers (including Fuller 1988b, ch. 9, and even Laudan 1986) have cast doubts on the wisdom of this approach, at least it paid lip service to the need for a theory of the scientific reasoner. In now separating the claim that science has made progress from the claim that science has proceeded by individually or collectively rational means, Laudan has not so much eliminated the problematic intuitions as relocated them to the more secure ground of that inveterate kibitzer, the reflective historian. What *has* been eliminated, however, is the scientific reasoner, who Laudan concedes is philosophically unfathomable. The alleged source of this unfathomability is relativism, i.e., scientists have differed so significantly over their ends and means that the philosopher cannot reconstruct a theory of scientific rationality without doing violence to the situated character of the reasoning of actual scientists. As will soon become evident, however, Laudan's interest in respecting the scientist's situatedness sits uneasily with the idea that the scientist and the historian partake of a common historical trajectory.

My overall appraisal of Laudan's strategy, assuming that it can be carried through, is that it gives up rationality much too quickly, seeing a trade off between rationality and progress where none need be seen. Rather, both a theory of the historian's reasoning (the "progress" account) and a theory of the scientist's reasoning (the "rationality" account) are needed. No doubt, Laudan is forced into perceiving a trade-off here because he holds to an overly subjective view of knowledge (i.e., as a set of rationally accepted beliefs), which most naturally lends itself to an image of the history of science as a succession of snapshots, with each snapshot depicting a scientific community at a particular time selecting a given theory for a given set of reasons. Missing here, of course, is any explanation of how it is possible for a theory to have been originally accepted by scientists for one set of reasons, yet subsequently be used by other scientists—those contemporaneous with the historian—for completely different epistemic ends (cf. Fuller 1988b, ch. 10). In short, Laudan—and analytic philosophers of science more generally—fail to account for our epistemic history as a process *during which* knowledge comes to be objectified. Why this obvious failure? Leaving aside the fact that much of the history of this process—focusing as it would on such institutional middlemen as gatekeepers, ideologues, instructors, and hacks—has yet to receive the sort of glamor that attracts research grants, there is probably a latent fear that close attention to the epistemic process would reveal that only some surface linguistic structures and gross behavioral patterns are in fact transmitted intact for any great length of time. (The therapy for this anxiety is social epistemology.)

On a more positive note, Laudan's efforts to distance progress from rationality perform something of a service in demystifying much of the Hegelian rhetoric associated with Lakatos' "rational reconstruction" of the history of science. Laudan's new project is called *normative naturalism*, and it makes the thrust of the Lakatosian project more evident by showing that the sort of judgments that Lakatos tried to tap by reconstructing what a scientist would have decided under ideal epistemic conditions (i.e., with regard to both evidence and cognitive aims) is best cashed out in terms of a historical meta-judgment: i.e., today's historian's evaluation of the impact that the scientist has had in facilitating the development of current science. Ultimately, however, this demystification is a mixed blessing. After all, we must not lose sight of the fact that Laudan's revised research program gives a new sense of legitimacy to precisely those features of the Lakatosian project—its Whiggish disregard of what the actual scientists thought— features that were originally considered most *objectionable*.

From what has been said so far, it looks as though Laudan's turn away from Lakatos and toward "naturalism" simply involves admitting that judgments of rationality are always embodied, i.e., made from a particular point in history, relative to particular aims, in this case epistemic ones. Very little follows from this alone about the validity of the judgments made, especially their range of applicability. Indeed, it may be that some periods are better than others in affording the sort of enlightenment needed for making universal judgments about the aims of history. As followers of the other great normative naturalists— Hegel, Marx, and Lakatos (Hacking 1981b)—are fond of noting, being later rather than earlier in the sequence of world-historic science confers the privilege of reflecting on one's predecessors, which better enables the successors to uncover the aims that have been guiding everyone throughout the process. However, it would seem that, by neatly parceling out the historian's hindsight into a theory of "progress," rather than what might less misleadingly be called a theory of "metarationality," Laudan tactfully sidesteps the less flattering side of his normative naturalism, namely, that the "rationality" he reserves for past scientists is nothing more than their successful pursuit of relatively immediate goals that bear only adventitiously on the overall aims of the history of science. In other words, the scientist's adaptability to her life-situation is inevitably treated as a symptom of her historical shortsightedness.

To be fair to Laudan, it should be said that a robust sense of relativism toward the past is difficult to maintain without also claiming that the historical situatedness of scientists from different periods makes their activities strongly incommensurable (cf. Taylor 1982). Otherwise, there is always the temptation to regress to a nineteenth century liberal grounding of one's "respect" for other cultures— including one's own past—in the understanding that, had the natives been capable of greater powers of mind, they would have displayed those capacities. As it stands, so this "liberal" argument goes, the natives managed to flourish in their habitat, which would seem to count against the reasonableness of holding them accountable to epistemic standards that they could not in principle meet! The ultimate source of the difficulty here goes to the very heart of "naturalism" as an epistemological posture, a study of which will be our first move toward retrieving rationality from Laudan. For all their substantial differences, the many varieties of naturalism share an underlying assumption: Disputes over judgments of value (the value in this case being rationality) are resolvable by appealing to the facts. Thus, our nineteenth century liberal based her condescending version of relativism on what she took to be facts about the intellectual limitations of the natives. What these "facts" are, and how one comes to learn of them, are, of course, the bone of contention among naturalists.

2. Shards of a Potted History of Naturalism

The classical philosophical tradition, which extends from Plato through Descartes to contemporary conceptual analysis, looks largely with disfavor upon naturalistic attempts to resolve value disputes, if only because naturalism seems saddled with the paradox that the very things that are supposed to resolve such disputes—the facts—are themselves contested objects, given to the vagaries of methodological disputes among the various academic disciplines, not to mention schools of empiricism. By contrast, classical philosophy has relied on *a priori* reasoning and, more recently, on that reified branch of ethnosemantics, "conceptual analysis," to resolve value disputes. These methods are said to proceed univocally from indubitable first principles (either universal truths or explicit definitions) to their logically deducible consequences. Left to her own devices, however, the naturalist would reinterpret what the classical philosopher is doing so as to render it equally contestable or at least contingent. Those "indubitable first principles" may be little more than ingrained "habits of the mind," as the American pragmatist Charles Sanders Peirce would have said. Their supposed "intuitiveness" merely reflects the fact that they are automatically brought to consciousness, whenever the philosophically appropriate cues are given. Indeed, latter-day attempts by ordinary language philosophers to bring a priori reasoning down to the earth by analyzing concepts, or "meanings," have openly courted naturalism. Although Ryle, Strawson, and Austin shied away from the empirical strictures of ecological psychology, all three regarded the domain of common sense bounded by ordinary language to represent the acquired wisdom of communities living through the human condition (e.g., Strawson 1959, p. 11). Perennial semantic distinctions, such as between persons and things, are thus explained in terms of the contribution that they make to the survival of the species.

However, to say that a carefully articulated account of a priori reasoning quickly veers into naturalistic psychology is to bring us no closer to what naturalism itself is. The tension inherent in the idea can already be sensed in Laudan's efforts to strike a balance between the historian's third-person perspective on the significance of some past scientist (i.e., the theory of progress) and the scientist's own first-person account (i.e., the theory of rationality). But the tension is perhaps even more keenly captured in the distinction between *naturwissenschaftlich* and *geisteswissenschaftlich* methodologies for the study of human beings, both of which have been touted as "naturalistic." In pursuit of the former approach, the following terms are associated with naturalism: *third-person, experimental, causal.* In pursuit of the latter approach, these terms are used: *first-person, ecological, phenomenological.* As a result, virtually every social science methodology can lay claim to being "naturalistic" in some accepted sense or other (which makes the rhetoric of naturalism a prime area of study [cf. Fiske & Shweder 1986]). Indeed, the semantics of the situation becomes especially vexed once a social scientist gets hold of Quine's (1969) "Epistemology Naturalized" and starts trying to make sense of the radical translation episode, in which a field linguist attempts to interpret native utterances from scratch. This episode epitomizes the ambiguous gestalt that is naturalism: If the social scientist focuses on the ecological setting of the task, it looks *geisteswissenschaftlich;* but if she focuses instead on the behaviorist orientation that the linguist adopts toward the natives, then its *naturwissenschaftlich* aspects become more evident.

Any potted history of naturalism must start with Aristotle's advocacy of a unified scientific method, one with a strong *geisteswissenschaftlich* bent. The subsequent history of naturalism tells of successive sciences establishing themselves on a more *naturwissenschaftlich* footing, especially as it became clear that there were aspects of nature that could not be comprehended within the perceptual realism associated with naturalistic observation. Special instruments and controls were therefore required to tease these hidden aspects out into the open. The most noted of such phenomena in Aristotle's day, magnetism, was the subject of early experiments by his student Straton, who went on to found the first great instrumentarium, the Museum at Alexandria in Egypt (Fuller & Gorman 1987). Moreover, it should not be forgotten that, true to the *Geist* in *Geisteswissenschaft*, Aristotle grounded his naturalistic inquiry in a search for the "animate" principle, the paradigm case of which was the organism persevering in face of resistance from its environment—perhaps the most obvious figure-ground relation that perceptual realism has to offer.

Indeed, as late as the first half of the nineteenth century, when biology could still be arguably classed as one of the "historical sciences," the German historicist Johann von Droysen may be found providing an Aristotelian account of what it means to "explain" and "understand" the member of any animal species, not merely Homo sapiens. Explanation and understanding (a methodological dichotomy introduced by von Droysen) are presented, in good Aristotelian fashion, as two temporal standpoints from which any animate being may be regarded: Explanation involves adopting the "external" perspective of past events, usually environmental ones, that determined the individual's current state, whereas understanding involves adopting the "internal" perspective of the ideal future state, or telos, toward which the individual is interpreted as developing (cf. Apel 1985, ch. 1).

During the second half of the nineteenth century, as John Stuart Mill's positivistically oriented *System of Logic* became the dominant philosophy of science in Europe, "naturalism" became a decidedly *naturwissenschaftlich* affair (cf. Lindenfeld 1980). While Wilhelm Dilthey continued to stress the ontological differences between *Natur* and *Geist* as grounding the need for two distinct forms of inquiry, most methodologists (especially such Neo-Kantians as Wilhelm Windelband, Heinrich Rickert, and Max Weber) regarded the issue in purely epistemological terms, specifically the radical empiricist terms that Mill had introduced. To wit, knowledge had two sources: direct acquaintance of particular events or inference from repeated experience to universals. Presented with just these two options, the only way to systematic knowledge was by repeated experience, since direct acquaintance is, by definition, based on precisely those features of experience that are most transitory. As Mill's inductive canons were meant to show, our natural tendency to accumulate repetitive experience in memory (which, in its natural state, is a cognitive liability based on our tendency to forget differences) can be refined into a reliable tool of inquiry through experimentation—a tool that could be applied to any domain, be it human or non-human. Thus, Wilhelm Wundt's self-consciously Millian experimental psychology demonstrated that one could have a *Naturwissenschaft* of human beings (Fuller 1983).

But as Wundt himself keenly realized, Mill's reformulations undercut the epistemological foundations of the humanities, which presupposed a much more Aristotelian sense of naturalism (cf. Tweney 1989). After all, humanists claimed to inquire into the meanings of particular documents without ever having had acquaintance with the author or even with the author's culture, a possibility that Mill's radical empiricism did not permit. The humanities have never really recovered from this epistemological crisis. The most adaptive response to the crisis has come in the form of ethnographic approaches to anthropology, where such traditionally humanistic modes of epistemic access as *Verstehen*, or "sympathetic understanding," are allowed literal application within an empiricist framework, given that anthropologists *do* have direct acquaintance with the particular people whose activities they are trying to interpret. By contrast, and again betraying the extent to which Mill still sets the terms of the discussion, *Verstehen* is nowadays regarded as only imaginatively or metaphorically applied (and hence to be supplemented, insofar as this is possible, by the standard inductive tests) in the more document-bound, historical branches of the humanities, even though *Verstehen* first arose in those disciplines.

3. Why Today's Naturalistic Philosophy of Science Is Modeled More on Aristotle than on Darwin

The legacy of Aristotelian naturalism lingers in the unit of analysis that is most commonly used to study people and animals in experimental psychology, namely, that creature of perceptual realism: the individual organism. Far from being a trivial metaphysical point, the continued focus on this individual has led in recent years to widespread skepticism about the feasibility of any normative attempts at improving human behavior, since *individual* human beings typically do poorly on experimental tasks designed to test their rationality. But this is to jump the gun somewhat in the discussion. We first need to confront the problem of individualism squarely and use the resources provided by the disciplinary cluster that has handled this problem in the most sophisticated fashion—evolutionary biology.

Ronald Giere (1988) is perhaps the most avid naturalist in the philosophy of science today. However, Giere is also a staunch individualist who explicitly blames the failure of traditional philosophical and sociological explanations of science on their abstracting "at too high a level" from the "causal locus" of scientific activity, namely, individual scientists (Giere 1989). Giere's "naturalistic" strategy is to study scientists in their laboratory habitats, watching how they manipulate apparatus and draw inferences from those manipulations. Giere's goal is to make sense of what the scientists are doing, and, as it turns out, what the scientists are doing generally makes the sort of sense they think it makes. This point would hardly be worth mentioning, except that Giere takes it to be an objective empirical finding on his part, rather than a relativist interpretive presumption. Herein lies Giere's nod to Aristotle. Science is whatever scientists happen to do in the environment to which they are best adapted, namely, the laboratory. And although Giere does not quite go so far as to privilege the scientists' own self-accounting procedures, he nevertheless accepts the frame of reference implied in those procedures. In other words, Giere trusts the scientists' own experience as an indicator of what causes them to act as they do. Thus, the fact that scientists talk about their work in realist-sounding terms is taken to be an argument that scientists are motivated by, among other things, realist epistemological considerations (cf. Leplin 1984b). (For the record, it should be noted that this is not the usual defense of scientific realism, which eschews scientists' self-reports, turning instead to the best third-person explanation for the entire track record of scientific achievement [e.g., Churchland 1979, Boyd 1984, Hooker 1987].)

In short, it would seem that Giere treats scientists as an Aristotelian natural kind, i.e., as members of a species that maintains a common form in the face of changes in the environment. Thus, Giere's naturalistic observer can spot the characteristically "scientific" way of responding to the range of events that occur in the laboratory. Indeed, Giere's most telling gesture to Aristotle is that like the members of an Aristotelian natural kind, Giere's scientists can *only* maintain their form but never change it substantially as a result of interacting with their environments. Despite the importance attached to historical case studies, Giere (1988), as opposed to Shapere (1984), does not countenance the possibility that the nature of science may itself radically change in the course of history.

From the standpoint of evolutionary biology, Giere's focus on the individual scientist as the object of his naturalistic inquiry equivocates on the meaning of "individual." To see this point, we need to do a little metaphysics. All metaphysicians agree that an individual is an entity that is bounded in space and time. Controversy arises, however, once we try to specify the "principle of individuation," in other words, what makes something an individual. Consider these two options:

(a) *Folk Individuals*: Entities are individuated from one another by perceptual discrimination, as in the case of our ability to distinguish two animals in a field because we see a space separating them. These may, but need not, be real individuals, depending on whether reality possesses the structure that we happen to perceive. Aristotle is generally taken to have believed that the two sorts of individuals are, in fact, the same, which follows from his view that humans were designed to understand reality.

(b) *Real Individuals*: Entities are individuated from one another by being different instantiations of the universal principles underlying reality. Depending on the nature of the principles, these individuals may, but need not, be perceivable as discrete entities. As explained below, evolutionary biologists regard an entire species as one real individual, even though we perceive a species as a sequence of discrete entities (i.e., successive generations of organisms).

Armed with this distinction, we can now identify two sorts of science:

(c) *Folk Science*: An inquiry that aims for empirical generalizations about folk individuals, on the assumption (based on an epistemology of perceptual realism) that these individuals are the real ones and that generalizations about them are the universal principles underlying reality.

(d) *Real Science*: An inquiry that aims for the universal principles underlying reality and the real individuals that instantiate them, on the assumption that folk individuals are probably only parts of real ones, and that there are unlikely to be any generalizations that range over folk individuals, or if there are such generalizations, they are probably restricted in scope.

As we shall now see, Giere's naturalistic inquiry, like other *geisteswissenschaftlich* enterprises, constitutes a "folk science" and as such is subject to the strictures of the methods of a "real science," such as evolutionary biology. I then argue that if we want to study science in the manner of a real science, then we must recognize that science, like an animal species, is not a natural kind, but a real individual: that is, an entity that has no essence of its own but is merely the product of a contingent interaction of real principles. In the case of science, these principles are to be derived from the social sciences.

The philosopher of biology David Hull (1974, p. 48) has observed that the perceptual realism of common sense embodies a particular ontology, namely, one in which the bearers of essential properties, or "loci of causal powers," are identified with whatever can be regularly perceived as discrete entities. This helps explain why pre-Darwinian biologists defined a species as the set of individuals that visually stand out from their habitats, in virtue of some palpably shared properties. Indeed, even much contemporary philosophical discussion about natural kinds falls back on this Aristotelian intuition, that the paradigm case of a thing that has essential properties is a member of an animal species (cf. Putnam 1975).

However, one of the major achievements of the Darwinian Revolution was to discredit this use of perceptual realism, by arguing that a species is only a conventionally defined set of organisms which, on the basis of phenomenal similarities, are thought to be products of a common genetic lineage. But if, say, two groups of these organisms do not exchange genes frequently enough, then, for all their surface similarity, they are not members of the same species (Hull 1988, ch. 3). In traditional metaphysical terms, this means that an entire species is itself best regarded as an individual composed of a subset of all "genes" (a term that I use as a placeholder for whatever turns out to be the "universals" out of which real biological individuals are composed). Thus, more like Leibnitz than Aristotle, the true Darwinian treats the members of a species as the partial realization of all possible forms rather than as the paradigmatic realization of one such form. (Of course, the very important difference between Leibnitz and the Darwinian is that the Darwinian does not believe that the "possible forms," or genes, have been a priori fixed. Rather, evolutionary history itself, the exchange of gene combinations through successive generations, causes changes in the set of genes that can be subsequently combined [cf. Hull 1983].)

What is the upshot of all this metaphysical maneuvering? In the first instance, it cautions the naturalist against confusing what the scholastics called "the order of knowing" (ordo cognoscendi) with "the order of being" (ordo essendi). If modern evolutionary biology has taught us anything, it has been that although individual organisms may be first in the order of knowing (i.e., it is with them that our biological inquiries begin), they are probably not first in the order of being (i.e., they are not the units into which biological reality is ultimately divided). To put the point more in Giere's terms, the *locus* of causal powers (i.e., individual organisms) must be distinguished from the causal powers themselves (i.e., genes). Evolutionary biologists are Aristotelian only at the start, noting that certain individual organisms survive long enough to reproduce in a given environment, whereas others do not. However, this observation represents only the grossest interaction effects of the mechanisms of variation, selection, and transmission that biologists are trying to understand. And these mechanisms, in turn, operate on different units of analysis. For example, even if one says that the environment selects individual organisms for survival, that says nothing about the traits transmitted from those organisms to their offspring that will allow the offspring to survive. It may turn out that traits that proved advantageous to the first generation do not serve the second one so well. The story becomes further complicated once we consider that the presence of other individual organisms constitutes part of the selection mechanism. In that case, some traits may have gained a selective advantage for their bearers simply because of the nature of the competition. Once the competition has been eliminated, the very same traits could turn out to be disadvantageous to the offspring receiving them. The story could be made even more complicated if one were to discuss the variation mechanism, and parse out "traits" into proper genic units (cf. Brandon & Burian 1984). However, from what has been said so far, it should be clear that once contemporary biologists try to sort out the causal mechanisms involved in evolution, the ontological status of the individual organism quickly fades to that of a convenient spatiotemporal marker for locating interactions among the relevant mechanisms.

4. Why a Truly Naturalistic Science of Science Might Just Do Away with Science

Now, how do these nuances of evolutionary biology affect naturalistic inquiry into science? First, as has befallen the terms that picked out natural kinds in Aristotle's biology, "science" (or, strictly speaking, "scientist") does not refer to some necessary combination of properties (i.e., an essence), but rather only to an historically persistent combination of such properties. This point would seem to have very important reflexive implications. For insofar as science searches for causal mechanisms that are expressed in laws, and insofar as these pertain only to necessary combinations of properties, then since science itself has no necessary properties, then there can be no laws governing science, and hence no "science of science," understood as a *naturwissenschaftlich* enterprise. The same line of reasoning applies to the possibility of there being a science of a particular animal species, such as human beings (cf. Rosenberg 1980). Arthur Fine (1986b) has made the point well: "Many sciences contribute to our understanding of the horse, but there is no 'science of the horse.' From an evolutionary point of view, there is only a natural history. I believe the same is true of science itself" (p. 175).

There are at least two responses to this argument. One would be to conclude (as Fine himself does) that it bolsters Giere's more *geisteswissenschaftlich* enterprise and ought to discourage those experimental psychologists and sociologists of knowledge who, in their very different ways, believe that only the self-mystification of scientists has impeded progress toward discovering the laws governing scientific behavior. The second response, which I take more seriously, is that we look toward developing a science of the relevant properties, some combination of which are recognized conventionally as scientific, and simply deny that there are any properties that are *essentially* scientific.

In short, I am calling for an *eliminative sociologism* with regard to science. Each property of science is already the subject matter of an existing social science, but the vicissitudes of disciplinary boundary maintenance in the social sciences have delayed the binding of the relevant parts of psychology, anthropology, geography, sociology, political science, economics, and linguistics into a unified "metascience." Such a metascience would be comparable to evolutionary biology (which is itself, after all, a cluster of disciplines ranging from population genetics to systems-ecology), except that the relevant properties here would clearly be, not of genes, but of basic behaviors. Let me spell out this last point in a little more detail.

Recall how Giere framed his perception of scientists, namely, as people engaged in a relatively self-contained activity located in a laboratory, involving what appears to be a unique set of skills. Certainly, the apparent uniqueness of scientists' behavior is reinforced by the sorts of environments they inhabit, whether we mean the special training that scientists must undergo or the special places in which they conduct their work. (And here we should keep in mind not only sensorimotor coordination but also linguistic behavior.) Moreover, this image of the relative autonomy of science is promoted by the scientists' own sense of class (or should I say, "guild") consciousness, which causes them to move through their careers in ways that approximate what Donald Campbell (1958) has called a "common fate."

However, taking a cue from B. F. Skinner (1954), we may justifiably wonder whether all this means that science is a truly unique activity, with an essence of its own or simply a matter of selectively reinforcing a concatenation of behaviors, each of which can probably be found selectively reinforced with other behaviors in other social practices. From an evolutionary standpoint, it would be crucial to show, not mere homologies between social practices, but the actual transmission of a behavior pattern from one social environment to the next. The trick, of course, is to individuate the selectively reinforceable behaviors in a given society, or the "social operants," as Skinner might call them, so as to demonstrate how they could be transmitted and integrated in a given set of environments to form a continuous social practice. (For a cross-cultural historical sociology that develops this perspective, see Runciman 1989.)

How far are we from the epistemic utopia of eliminative sociologism? Distance here should be measured in rhetorical terms.

Talk of science as, say, "inherently rational" causes us to react to the effects of scientific research in ways quite unlike the way we would react to similar effects that are attributed to non-scientific sources. The antidote, then, is a comprehensive demonstration that science and society are *interpenetrative*: one always implicates the other. Luckily, the most avant garde work in Science & Technology Studies today, actor–network theory (Callon, Law & Rip 1986, as popularized in Latour 1987a), has already begun this task. The most rhetorically effective way of continuing it is by a two-part history of science that stresses the combinability of strands of a scientific culture with other strands from outside that culture:

(A) The first part of this historical project would deal with the transfer of skills from non-scientific (or other-scientific) sectors of society to the scientific sector under study. This would typically happen in one of two ways, as illustrated in the following example: either someone trained in engineering would deploy the mathematics she was taught on problems in economics or a trained economist would turn to engineering mathematics for some analogies that might be useful in solving standing problems in economics (Mirowski 1989).

(B) The second part of the history would involve the somewhat more innovative strategy of locating production sources common to scientific and non-scientific spheres of society. The easiest way to think about this possibility is in terms of a company that manufactures capital goods, such as computers and meters, that can be used in either scientific or non-scientific settings. How do these goods need to be "adapted" to their settings in order to be defined as producers of "knowledge" rather than, say, of some other material good? Histories of information technologies, which typically attribute a large causal role to big business (e.g., Beniger 1987), would be a good place start looking for clues. Another lead would be to study scientists who have managed to parlay government-sponsored, problem-oriented research (usually for national defense) into self-sustaining fields of inquiry severed from the interests of their original sponsors. Much of the post-World War II research on machine learning, experimental psychology, and cognitive science was initiated in just this fashion (Galison 1993, ch. 7).

I expect that a given science consists of patterns of labor organization, motivational and power structure, communication, codification, and apparatus manipulation that can be found in other, normally unrelated spheres of society, but that are made relevant to one another precisely by all these behaviors being regularly (and after a while, mutually) reinforced in a common environment. The ease with which one such behavior pattern can elicit the others in the appropriate environment ultimately leads scientists and their more credulous observers to conclude that there is a common "content" of which all these behaviors are convergent indicators. Thus, talk starts to abound to the effect that, say, "chemical knowledge" is embodied equally in the chemist's journal scribblings and in her adroit handling of test tubes. At that point, chemical knowledge becomes a proper object in its own right, even though it is little more than a creature of perceived coincidence, a Baconian Idol of the Theatre. Precedent for my line of reasoning here can be found among recent psychologists (e.g., Skinner 1957), anthropologists (e.g., Harris 1963), and sociologists (e.g., Bourdieu 1981).

Needless to say, just as the selective reinforcement of certain social operants maintains the existence of science, the absence of such reinforcement would spell the end of science (cf. Foucault 1970, on "the death of man"). Indeed, from the standpoint of breaking down the image of scientist as natural kind, perhaps the most crucial piece of the picture that I have been sketching is that each social operant has its own dynamic tendencies that are independent of the operant's role in a deliberately maintained social practice. Worthy philosophical prototypes for my view include Hume's "habit," Max Weber's "tradition," and Sartre's "practico-inert," all of which were designed to account for characteristically human activity without having to postulate intentionality. My basic claim here is that behavior patterns change in explicable ways, simply in the course of being reproduced from one environment to the next, and in conjunction with other behaviors. Moreover, one can specify these changes largely without having to make reference to the reasons why particular people would want to engage in these behaviors at particular points. The most obvious cases of this phenomenon concern the integration of formerly discrete behaviors into a fluid practice, as routinely happens to scientists who, in the course of becoming habituated to a new experimental technique, no longer need to interrupt the flow of their lab activities by consulting a written account of the technique.

Since historians of science have typically been preoccupied with the work of geniuses rather than of more statistically representative inquirers, there is a remarkable paucity of data on the frequency and distribution of the various behavior patterns associated with science (Shadish & Neimeyer 1989). Thus, although much may be known about what a given scientist did in her lab and how she accounted for it, relatively little is known about whether her behavior is representative of what other similarly trained scientists would have done in her situation. It is not enough to suppose that because the scientist managed to pass the scrutiny of her peers that her behavior is representative of what they would have done. As I will show in Chapter Four, the conditions under which one imposes sanctions on another's behavior are not necessarily a good predictor of what one would do in a similar situation or even of what one would say she would do in such a situation.

One of the most important and well-documented, yet least palpable, changes to result from the sheer reproduction of behavior in a variety of times and places is *semantic drift*. Words tend to accumulate meanings as the number of contexts in which they are used increases. However, our ability to monitor this semantic elaboration, as well as our ability even to discriminate from among the elaborated meanings, is limited. Consequently, instead of our verbal behavior's becoming indefinitely more nuanced, once we reach the limits of our ability to distinguish meanings, the different meanings start to be associated with one another, simply because they are attached to the same word. And although this tendency of homonymous (or homophonic) words to collapse into synonyms serves to streamline the communication process, it also tends to ghettoize inquiry into large amorphous problem areas, such as "the Nature–Nurture Controversy," whose depth and intricacy are an artifact of our inability to keep separate many superficially related issues for any considerable length of time.

There are three points to notice about semantic drift as illustrative of the dynamics of social operants:

(1) Behavior patterns tend toward maximum economy, in part as a function of the performer's having become more practiced in the behavior, and in part as a function of the audience's being limited in its ability to discriminate responses that are licensed by the behavior. If this part of the story is true, then we should expect to find many cases in the history of science where someone had worked out a detailed solution to a standing problem, but because the solution was so difficult to follow (due to the length, complexity, or idiosyncracy of the account), it was virtually ignored, until someone else arrived at a simpler—and perhaps even less adequate—solution.

(2) The major problem areas of a field, upon which inquirers deliberately focus their efforts, are essentially the unintended consequences of the economization of behavior patterns: that

is, a product, not of conceptual design, but of a failure to regulate behavior effectively.

(3) Any deliberately maintained social practice, such as a science, is a relatively unstable cluster of behavior patterns, which, if not closely monitored, will unravel according to the dynamics of each of the relevant social operants. This last point will be elaborated later when I account for the disparate tendencies in the four behavioral modalities in which norms are expressed in science. In brief, I mean here, among other things, the unnerving tendency for scientists' discourse to be reinforced independently of their other sensorimotor behaviors, and hence (pace Giere) to be a poor predictor of those behaviors.

5. A Parting Shot at Misguided Naturalism: Piecemeal Approaches to Scientific Change

For a parting shot at how the terms of naturalism can be radically transformed by taking seriously the metaphysics underlying evolutionary biology, let us briefly consider the highly vexed issue of scientific change, where a scholastic search for the middle ground often impersonates a dialectical synthesis.

The first moment of this would-be dialectic was when the positivists suggested that scientific knowledge is distinguished by some metric of continuous growth, such as the steady accumulation of facts or the subsumption of more phenomena under fewer laws. The second moment was provided by Thomas Kuhn, who claimed that the most important changes in science were radically discontinuous, indeed a revolutionary transition between incommensurable paradigms. The putative synthesis, then, is to say that there are some local discontinuities between successive paradigms, but that these are not incompatible with net epistemic growth in the long run. At this point, we are treated to some enchanting metaphors. First, Quine (1953) painted the picture of Neurath's boat, whose planks are successfully replaced one by one while at sea. More recently, Laudan (1984) has spoken of "reticulation" and the historian of physics Peter Galison (1987) of "intercalation" as models of scientific change. Both images contribute to the idea that the body of scientific knowledge is an inert creature broken up into parts of just the right size to be taken up for separate inspection. According to these "piecemealists," one such part, say, a theory or a method, can be examined, hotly contested, and even substantially changed, while the rest of the scientific corpus remains intact in the "background." Indeed, as Shapere (1987) and Laudan have emphasized, this very fact—that agreement is presumed over the rest of the knowledge base for the sake of arguing about the epistemic status of a given part—makes the emergence and resolution of disagreement in science such an efficient, and in that sense "rational," process. But now consider two presuppositions of this viewpoint:

(e) No part of the knowledge base can change, unless it has first been formally introduced for consideration. To hear Shapere tell it, one would think that everything in the current scientific corpus had to have been explicitly added at some point in history, and if the scientific community sees fit, may be explicitly removed in the future.

(f) The parts of the knowledge base that are placed in the background for purposes of contained disagreement do not change during the disagreement. Since whatever changes in the knowledge base must change explicitly, what is left implicit, such as background agreement, does not change.

These presuppositions help give philosophical accounts of the history of science their overly self-conscious quality, as if one were in the midst of parliamentary debate, with all the members keenly aware of the motion on the floor, since they collectively decided to debate and, ultimately, to resolve the motion in one way or another. Hence, in the hands of philosophers, the history of science becomes a history of "theory choices," with the philosopher the one who determines whether the rules (of rationality) were followed in making the choice or whether some formally prohibited issues (e.g., social interests) were illicitly insinuated into the deliberations. In any case, the philosopher's scientists always pay attention to each other and never seem to be daunted by differences of time, space, and culture.

An adequate explanation of why philosophical discussions of thinking and doing always seem to turn into accounts of debating and voting (needless to say, the tendency is even more pronounced in moral and social philosophy than in epistemology) would require nothing short of a psychoanalytic interpretation of the philosophical Ur-myth, debate among peers in the Athenian polis. But for our purposes, a less primal diagnosis will do. Whatever their view about the natural world, philosophers are inveterate antirealists about the social world, largely because—so it would seem—they falsely assume that we exert more cognitive and practical control over effects we cause than over those that we do not (Fuller [1988a] calls this "Vico's Fallacy"). Moreover, the recent piecemeal approaches to scientific change court a rather strong version of social antirealism, for these views suggest that no change has been brought about in the scientific corpus until the change has been recognized as such by the scientific community; hence, Laudan's (1984) consensualist theory of validation. But the piecemealists are certainly not alone in these assumptions. In this respect, we should be alerted to a curious, though decidedly underplayed, metaphysical allegiance between the piecemealists and their main foes, the social constructivists in Science & Technology Studies. Neither group believes in the independent reality of the social world, which explains why each in its own way is oblivious to the unintended consequences of human action, the alienation of reason from the reasoner, and the latent functions of apparently irrational practices—all staples of macrosociological theory of every possible ideological stripe, ranging from classical political economy, through Marx and Durkheim, down to contemporary American, French, British, and German structural-functionalism (Outhwaite 1983; Alexander et al. 1986). Indeed, as far as I can see, the only reason why a piecemealist like Laudan does not believe in an instantaneous switch in paradigms, whereas a social constructivist like Harry Collins (1981) might, is that the piecemealist presumes that a great many more people need to change their minds before the revolutionary shift has occurred, not merely the "core–set" of a research team who turn out to be the vanguard of change. That is, as a matter of logistics, the piecemealist needs to draw out the duration of the revolution in order to achieve the larger consensus.

Aside from the aesthetic appeal (such as it is) of moderating two extreme positions in the philosophy of science, what are the attractions of going piecemeal? According to Laudan (1984, p. 86), the main one seems to be that it disentangles several issues fused together in Kuhn's account of scientific change. In particular, Laudan means to deny these two Kuhnian moves: (1) if a scientific change occurs at many levels (i.e., involving changes in theory, methods, and even the aims of doing science), then all levels of the change must occur simultaneously (as in Kuhn's talk of "conversion" and "Gestalt switch"); (2) if a change is radical (as in Kuhn's talk of "revolution"), then it must occur all at once. Laudan is undoubtedly correct that the conclusion does not follow from the premise in each case, but that is not sufficient to license a piecemeal approach. After all, one key Kuhnism that is missing from Laudan's critique is the idea that scientific revolutions are *invisible* (Kuhn 1970a, ch. 11)—a term that, at the very least, suggests that revolutions can happen without the revolutionaries fully realizing the import of their acts, a point that the social antirealism of the piecemeal approach does not allow. In other words, Kuhn would seem to be a social realist of sorts. But rather than venturing into the thickets of Kuhn exegesis, let me now indicate where there might be a place for the piecemeal approach in a world where scientific revolutions can happen invisibly.

On the basis of recent social psychology research into the relation between memory and personal identity (Ross & McFarland 1988), I am persuaded of the following two theses:

(g) The principals in a scientific revolution come to understand the significance of their acts only in a piecemeal fashion over a great length of time. In fact, it probably takes the next generation or an objective historian to draw out all the revolutionary consequences.

(h) Revolutionary change is made psychologically possible by the fact that the principals notice the revolution only in a piecemeal fashion, since it is not clear that a revolutionary's sense of self could withstand knowing that so many changes are happening at once—assuming that the revolutionary had the cognitive capacity to monitor them all.

Thus, I grant a lot to the piecemeal approach, when it comes to the *phenomenology* of scientific change. But that just accounts for how scientific revolutions are perceived, not how they actually occur. My own view on this matter is to take the invisibility of revolutions seriously and reconceptualize the object of scientific change. As Fuller (1988b, ch. 4) argued, what changes is the relative burden of proof that positions must bear in the exchange of knowledge claims. This happens, as it were, subliminally, so that by the time the change is fully realized, it has become irreversible: For example, textbooks have canonized the new balance of epistemic power. Let me sketch how I envisage the steps of this process and then suggest how debates in evolutionary biology might provide conceptual aid.

The first movement in shifting the burden of proof occurs when an article or book passes through a discipline's gatekeepers so as to allow it maximum exposure. Perhaps the gatekeepers read the text as advancing current research, but unbeknownst to them (and perhaps even to the author herself), the text also rearranged the meanings of a few terms as well as restating some old arguments and claims in ways that made them appear more or less plausible than before. However, had any of these dialectical maneuvers been made the centerpiece of the text, the gatekeepers would probably have rejected it. Indeed, they may have previously rejected texts that highlighted such moves. But now, because the author has succeeded by the normal disciplinary criteria, the rest of the text is tolerated or simply ignored. Nevertheless, the precedent has been set for subsequent authors to open up the arena for dissent by drawing opportunistically on the legitimacy of the subtle semantic and rhetorical shifts. At first, most of these efforts may fail to get through the gatekeepers, but enough pass to foster a climate of dissent in the periphery of the discipline's consciousness. If sustained long enough, this tolerance may well evolve into tacit approval, and even explicit endorsement by the discipline. Part of this reversal of probative burden would be explicable in terms of a robust social psychological finding, *the sleeper effect*, whereby prolonged exposure to an opinion alone tends to make subjects more receptive to it, even if they originally attached little credibility to the source of the opinion (Hovland et al. 1965, ch. 6). It is as if familiarity breeds assent.

But human suggestibility is not the entire story (though it might end up explaining quite a lot, if Kornblith [1987] is to be believed). In addition, we need to take into account the change in disciplinary personnel across generations. A discipline's research program may have a long history of success undergirded by deep conceptual considerations. However, the entirety of this tradition is unlikely to be known by anyone other than a historian specializing in the area. Consequently, the discipline must continually redefine itself by mobilizing some fairly local historical resources, namely, the most recent books and journals. This point can work against an established tradition, if it politely ignores, yet tolerates, a minority voice that repeatedly defends its position in terms that have been subtly shifted to its advantage. In that case, the current generation of historically amnesiac scientists will probably see the debate in the minority's terms, precisely because the minority's are the only ones being used at the moment. Not surprisingly, this will serve to make the established tradition appear ungrounded, thereby placing the burden of proof squarely on the tradition's shoulders (cf. Noelle-Neumann 1984).

A good model for thinking through the complexities of scientific change, which resonates with the sorts of considerations that I have been raising against the piecemealists, is the turn of the century debates over the nature of individual variation operating in evolution. In arguing about whether evolution was "gradual" or "saltative," all sides seem to have confused the size of the mutation with the abruptness of its appearance. Such a confusion was possible because a given genotype (i.e., genetic blueprint) can be reproduced as many phenotypes (i.e., surface traits), each of which could also have been produced by any of a variety of other genotypes. However, natural selection occurs at the level of the phenotype, which means that the genotypic motor of evolutionary change is affected only indirectly. Thus, a mutation could be large but appear only gradually if an organism with a radically different genotype is able to transmit its genes, by virtue of manifesting a phenotype that is compatible with the other phenotypes selected in that environment. Thus, a major genetic change could transpire below an imperceptible change in surface traits. Darwin seems to have countenanced this possibility (Hull 1974, ch. 2). The above elaboration of Kuhn's talk of invisible scientific revolutions makes an analogous point, to which the piecemealists have so far been oblivious.

6. Towards a New Dismal Science of Science: A First Look at the Experimental Study of Scientific Reasoning

Recent experiments in cognitive psychology seem to show that no available theory of rationality has a basis in psychological fact. Consider this list of liabilities (cf. Ross 1977): subjects confirm when they should falsify; they ignore the base rate probabilities essential for Bayesian inference; they fail to see how sample size affects statistical reasoning in general; they cannot conceptualize causes whose interaction brings about an effect; they erroneously take the ease with which they remember something as an indicator of the extent to which it represents their experience; they do not make consistent expected utility assignments; and so on. A point that often goes unnoticed in the recital of this litany is that our cognitive fallibilities are sufficiently deep to cut equally against standard internalist and externalist models of rationality. In other words, not only are the "philosophical" accounts affected, but so too are the "sociological" accounts that portray the scientist as making key epistemic decisions on the basis of political or economic interest, for the experiments seem to show that human judgment is no better when applied to self-centered matters (e.g., weighting personal utilities) than when applied to more self-detached ones (e.g., weighting hypothesis probabilities). In fact, one of the more robust findings is that there is a natural asymmetry in the way people explain their own behavior vis-à-vis the behavior of others, an asymmetry that, although it serves the cause of self-vindication, also contributes to a skewed sense of the causal structure of the social world (Nisbett & Wilson 1977).

    Therefore, we should think of these experiments as potentially having global import, impugning any ability that we might claim to follow rules systematically or to reason self-consistently. Two of the leading cognitive psychologists, Amos Tversky and Daniel Kahneman, have come close to making this claim explicit in their demonstration of "frame–invariance violations," to wit, that even subjects who are able to make rational risk assessments in canonical cases fail to be equally rational when considering relevantly similar cases that are described, or "framed," in a non-canonical fashion (Tversky & Kahneman 1987).

Alvin Goldman (1986) is one of the few philosophers who has tried to relocate epistemology, now called "primary epistemics," in a world where the results of these experiments must be taken more or less as given. But as for Goldman's colleagues, all one can say is never underestimate the philosopher's ability to plant skeptical doubts. Given many of the considerations that have so far been raised in this book, one might think that the most obvious source of doubt would be the image of the rational agent presupposed in these experiments, namely, a strongly inertialist image, in which the subject must solve a problem alone with the barest of tools, props, and hence, cues in her immediate vicinity. The critique, in that case, would be that the experimental environment is too impoverished to permit the manifestation of rationality: Instead, you need several people working together for some time with a rich array of "reason-enhancing devices" (e.g., computers, scientific instruments) at their disposal. If the individuals have been organized in the right way (something that can be experimentally tested, cf. Whitley 1986), rationality will then emerge as the collective product of their endeavors. I will shortly return to this critique; however, it is *not* the one that philosophers have tended to raise. For an initial sense of the typical philosophical response, consider the inductive logician L. Jonathan Cohen (1986), who has been the keenest critic of these alleged empirical refutations of rationality.

Cohen essentially believes that, in most cases, subjects are deceived by the ambiguous tasks set in the experiments, which do not permit the periodic feedback that is normally necessary for people to become self-conscious of the rules of right reasoning. By contrast, education in logic, statistics, and the methods of the special sciences expose the student to the sorts of situations that one is likely to run across in her professional life. Although the textbook situations are themselves rather stylized, even they contain much more of the detail needed for prompting the deployment of the relevant rule than does the average experimental protocol. Thus, not only is Cohen, in our terms, an inertialist, but he also seems to think that we are *really* epistemic engines, already programmed with rationality and simply in need of some experience to sort out when the various principles are appropriately applied.

Whatever we make of Cohen's version of the epistemic engine thesis, it is clear that philosophers have generally suspected that the psychological findings are invalidated on the grounds that the experimental subjects are usually novices rather than experts in the specific reasoning tasks. This is somewhat ironic, given that the experiment most often used to show that people display a "confirmation bias" (i.e., they accumulate positive instances for a pet hypothesis but they never try to falsify the hypothesis) sets subjects a task that was inspired by philosophers (Popper [1959] and Wittgenstein [1958]) interested in trying to make some rather general points about reasoning. In the late 1950s, Peter Wason began modern research into cognitive limitations by asking subjects to guess the rule governing a series of numbers (e.g., 2, 4, 6, . . .) by having them propose other ordered triples (e.g., 8, 10, 12) to see whether or not they too conform to the hidden rule (Tweney et al. 1981, ch. 16). Needless to say, an experiment of this sort would be simple enough to perform on undergraduates without scientific training. Indeed, it may even be that because novices are used in these experiments, the tasks are devoid of specialist content. But this, in turn, opens the tasks to multiple interpretations by the subject, perhaps unbeknownst to the psychologist herself. The philosophical verdict, then, is that the experiments exaggerate the extent of our irrationality by ignoring the possibility that training may improve performance. In fact, Tversky and Kahneman themselves seem to grant this point even with regard to less disciplined forms of knowledge: If subjects are as incompetent in calculating expected utility as their experiments suggest, then success in the business world must be intimately tied to learning from one's mistakes over time—the so-called School of Hard Knocks (cf. Shweder 1987; cf. Pitt 1988).

Unfortunately for Cohen, and perhaps even for Tversky and Kahneman, while performance in reasoning skills does improve with training, it improves for the sorts of tasks in which one has been trained and only spottily beyond them. Studies of expert judgment across a wide variety of fields show that even in the cases where the experts surpass novices, the difference is not great enough to inspire confidence in the efficacy of prolonged training (Arkes & Hammond 1986). Moreover, laboratory ethnographers have repeatedly observed that, Cohen's complaints to the contrary, even real scientific problems are given to ambiguities of the magnitude found in the psychology experiments. Indeed, such basic identity questions as "Which problem was solved?" are settled only when the results are being written up for publication (Latour & Woolgar 1979, Knorr-Cetina 1980, Gilbert & Mulkay 1984). Consequently, even assuming that a scientist can be trained well in certain reasoning skills, that will hardly ensure success in the research environments that the scientist is likely to face. As for the less-structured environments faced by businesspeople, the chances for cognitive improvement are even murkier, since continued survival in the marketplace does not necessarily imply a more finely honed competence in economic matters. In most cases, survival has less to do with learning the means best able to attain one's ends than with learning to cope with the outcomes of one's efforts. Unlike the former skill, which would indeed draw on the businessperson's economic expertise, the latter is grounded more in the ability to adapt creatively to one's situation, often to aim for more but settle for less, or at least for something other than what was originally intended. In fact, creative adaptation may be possible precisely because our cognitive apparatus is sufficiently faulty for us to naturally forget, or otherwise ignore, the discrepancy between what was intended and what ends up happening (cf. Elster 1984b). The point may apply more generally to all attempts to learn from mistakes. After all, usually our only evidence for having learned something is that we henceforth proceed more easily upon a course that had previously been pursued only in the face of some resistance (Brehmer 1986, Hogarth 1986). But this is not the place to conduct a proper inquiry into the mythical status of learning from mistakes (cf. Fuller 1987a).

7. Sociologists versus Psychologists, and a Resolution via Social Epistemology

For a point of contrast with Cohen, let me now turn to a more sociologically inspired critique of experiments that purport to demonstrate our irrationality. I will focus on one experiment (Mynatt et al. 1978), in which undergraduates are asked to hypothesize about the laws governing a computer-simulated universe. Unlike the more cited and contested experiments, this one actually engages the subjects in a specifically scientific task; in addition, it departs somewhat from the usual practice of placing the subjects in Cartesian isolation in that this experiment has them interact with a video display terminal (especially to gauge the effects of shooting particles at objects in the universe) and allows them to keep track of their hypotheses in writing (this also lets the experimenter track their implicit methodology). The experiment was designed so that half of the subjects were instructed in the method of "strong inference" (i.e., testing many hypotheses at once, with an eye toward falsifying most of them) and half were not. Did the subjects trained in strong inference apply it when generating and testing hypotheses? If so, did they come closer to the right set of laws than the control group?

As would be expected, subjects did poorly, regardless of training. Indeed, the majority exhibited confirmation bias, the tendency that strong inference is designed to counteract. Moreover, subjects who tried to apply strong inference tended to falsify too many hypotheses on the basis of too little evidence, which left them doubting that the simulated universe was governed by any laws at all. This case is typical of the class of experiments under consideration in that if a person is of a particularly skeptical turn of mind, then she might conclude not only that people are incorrigibly irrational, but also that training them in the ways of rational methodology might make matters worse!

The sociological critique of this experiment has five parts:

(1) Given the artificiality of their research environment (geometric shapes moving on a video screen), combined with impoverished theoretical and technical resources, the subjects were confined to a strategy of predicting the motions of objects under various conditions. In other words, the subjects were locked into an instrumentalist methodology by not being provided with the cues that normally prompt scientists to hypothesize beyond the immediate evidence. Such cues could come from alternative modes of access to the simulated universe, which might suggest convergent ways of identifying common causal mechanisms. Thinking in these realist terms, where strong inference is usually embedded, would discourage subjects from treating all the evidence as equally revelatory, and hence dislodge the conservative hypothesizing policy of "saving the phenomena," the likely source of the subjects' confirmation bias.

(2) Since the "scientific activity" in which the subjects are engaged is little more than a slowed down version of a video game, there is hardly any point to doing well. Admittedly, the subjects were paid a nominal fee for their time and if they discovered the laws, a bonus as well. Still, these rewards are not intrinsic to the scientific process itself; in fact, they are clearly compensation for a task that the experimenter believes the subjects will find tedious. Thus, the experiment may be criticized for failing to simulate the motivational structure of scientists.

(3) It is not clear that the experimenter was very savvy in her efforts to teach the subjects strong inference. As it turns out, she gave them an article to read on the topic. However, I will argue later in the next chapter that the scientist's knowledge of methodological norms may be encapsulated to cover only situations resembling the one in which the norm was learned. Thus, studying an article on strong inference may enable the

subject to respond appropriately when asked about the sorts of hypotheses that scientists should propose. But there is little reason to think that this ability will automatically transfer to the subject's own scientific practice, unless she is specifically trained to embody the strong inference strategy in her research. Notice that, in a crucial sense, my critique is more radical than Cohen's related one. As I noted above, Cohen finds the learning environment of the experiments too impoverished to cue the right pre-wired normative responses. By contrast, I am claiming that the normative responses are themselves constructed out of the learning environment and permanently bear that mark of origin, which, in turn, explains the need for the experimenter to control how the subjects are trained.

(4) In this experiment, subjects were monitored for the hypotheses they generated and tested, but the experimenter did not give them any feedback unless they stumbled upon the right set of laws. Subjects adjusted their hypotheses entirely on the basis of how they interpreted the universe's responses to their probes. Notice that this is virtually the exact opposite of the scientist's real epistemic situation: The scientist, when interpreting an experimental result, normally consults with colleagues, but clearly "Nature" would not intervene to end the scientist's inquiries, were she to propose the right hypothesis. Admittedly, experiments of this sort differ on the amount of feedback given to subjects, but in all cases, the experimenter already knows the right answer and often informs the subject when she hits upon it. Thus, these experiments fail to simulate the socially negotiated, open-ended character of scientific inquiry.

(5) Another sign of the experimenter's absolute epistemic authority over the subjects is that the experimenter presumes that if a subject's pattern of hypothesizing does not conform to a strong inference strategy, then the subject is performing subrationally, given the opportunity that the subject originally had to learn the relevance of the strategy to the task at hand. However, not only might the subject not have been originally exposed to the ideal learning conditions (see [3]), but she may also be invoking principles of "metarationality," which warrant deviation from the principles of rationality, when deviation would seem to expedite a solution (cf. Jungermann 1986). This is apparently what the subjects themselves claimed during the debriefing session after the experiment.

And although subjects may simply be saving face when they reinterpret subrational behavior as metarational, nevertheless their doing so suggests that, at some point during the experiment, the subject's state of knowledge may have warranted the violation of the strong inference strategy. Since the subject was, in effect, taking a calculated risk, the fact that her shortcut did not work does not necessarily diminish its rationality as a strategy at the time it was chosen. However, this is merely a speculative defense of the subject's rationality, which would ultimately need to be cashed out with an account of the course that the subject saw her inquiry taking at various points in the experiment. In other words, the experimenter would need to conceptualize the subject as engaged in an historical process.

Lest the reader think that the relation between philosophers, psychologists, and sociologists of science must be entirely adversarial, I will now sketch an experimental research program that the psychologist Michael Gorman and I are developing to study aspects of knowledge transmission that are of particular concern to social epistemology, but that so far have been been given little empirical treatment.

The experiments are designed to be run on groups, each with four members, each of whom is replaced one by one, until the entire original group has been replaced. The model for this design is Jacobs and Campbell (1961), which examined the extent to which a false viewpoint (a so-called arbitrary tradition) can be transmitted intact as members of a group are gradually replaced. Do new members simply conform to the reigning viewpoint, or does an independent assessment of the situation cause them to change their mind. The false viewpoint in question was the autokinetic effect (i.e., a point of light projected in a dark room can be falsely perceived as moving in certain ways). Jacobs and Campbell found that the false viewpoint was perpetuated in an eroded form (i.e., the last generation of the group did not see the light moving as much as the earlier ones), suggesting that reality does slowly make some headway into collective illusion. Jacobs and Campbell argued that the false viewpoint was not more robustly transmitted because there was no social function served by holding it, unlike the superstitions that persist in real societies.

Instead of asking subjects to track a point of light in a dark room, we plan to get them to perform a complicated version of Wason's 2-4-6 task, in which subjects must not only guess the rule governing the number series but take in account the two colors in which the ordered triples may be written, as well as a letter of the alphabet that is appended to each triple. Gorman and his associates have already conducted several experiments using this task, which he has found engages the subjects long enough to elicit a rich set of problem-solving protocols (Gorman et al. 1984). The trick is for the experimenter to select a rule that is very general so that competing hypotheses can be constructed and maintained for a while.

This research program is designed to focus on three sorts of independent variables. The first is, so to speak, the *politics* of the problem-solving task. Specifically, we plan to consider three ways in which groups may interact while trying to discover the rule, adapted from Anatol Rapoport's (1980) three models of politics as a rational activity. Thus, assuming that two groups are proposing alternative hypotheses to discover the rule, each group may also be told that its aim is (A) to defeat the other group in successive rounds of hypothesis evaluation (*voting*), (B) to persuade the other group to accept its hypothesis (*marketing*), or (C) to eliminate the other group by arriving at the best hypothesis (*fighting*).

In (A), the two groups are competing like opposing parties in an election, with periodic evaluations of their hypotheses (e.g., who has gotten closer to identifying the rule) by the experimenter as objective third party. There is no communication between the groups while working on the task, and each member remains aligned with her original group. This situation is akin to the internalist historian's model of theory choice in science.

In (B), the two groups monitor their own success by whom they are able to persuade to their respective sides. The idea here is to get members of the other side to work on one's own project, which captures Pierre Bourdieu's "cycles of credibility" that supposedly determine the course of scientific research in the short term (cf. Latour & Woolgar 1979, ch. 5). However, the group with the smaller number of members may continue indefinitely, ever hopeful of converting the rest. It would be interesting to see whether the two groups, out of their own accord, come to agree that they have reached the same hypothesis; or whether the politics of the situation would perpetuate perceived differences, even in cases where the experimenter would say that they had reached the same hypothesis.

In (C), the two groups know that periodically one group will be eliminated. They can communicate, if they want, but the experimenter decides which group's hypothesis is closer to the rule, regardless of how many members the group has. Once a group has been eliminated, the remaining group may splinter off to form two new ones. This perhaps models radical conceptual changes in science.

A second sort of independent variable is the instructions that the subjects are given. In the sociological critique of Mynatt's experiment using the computer simulated universe, objections (1), (3), and (4) pertained specifically to the instructions. Subjects may or may not be instructed to adopt certain heuristics when looking for the rule. If no instructions are given at the outset, how and which heuristics emerge, in which sorts of groups, and how are they transmitted to new group members? Also, the instructions may be designed to give an "instrumentalist" or "realist" spin to the task: Are the subjects told to save the data or to look for underlying, even interacting principles responsible for the data? Do realist instructions better elicit a falsification strategy, and does that hasten a correct solution? How are these instructions then transmitted to new group members? In addition, groups may or may not be told that they are continuing a (perhaps fictitious) previous group's work. If they are told, and they are given the last hypothesis tested by that group, do they tend to triangulate around it—even if the hypothesis is far off the actual rule? Finally, Gorman (1989) has introduced an innovation designed to meet the objection raised by the sociologist in (4). Gorman alternatively tells subjects that there is actual random error, possible random error, or no error in the data that they receive from the experimenter after proposing an ordered triple as a test case for some hypothesis they have been entertaining. How is a group's behavior affected by its members' being told that they are possibly receiving enough error in the experimenter's responses to make their data unreliable? Do they, in some sense, "ignore" the data, and persist with pet hypotheses indefinitely? And if so, would this be sign of the subjects' constructivist tendencies or of their realist ones, since a realist might equally be inclined to think that the appearances are only a partial indicator of the underlying reality. An interesting case here would be one where subjects are told that there may be a large amount of error in the data, but in fact, unbeknownst to them, there is none.

The last sort of independent variable concerns group replacement. Groups may choose their own members, according to the cognitive resources they perceive themselves having and needing; or groups may be assigned new members by lot, or perhaps even by weighted lot, reflecting the duration of one's membership in the group (i.e., a member's chances of leaving increase with duration).

8. If People Are Irrational, Then Maybe Knowledge Needs to Be Beefed Up

If the various canons of deductive, inductive, and domain-specific reasoning turn out to be as psychologically unfeasible as they seem to be, does it follow that rationality has been empirically falsified? As we have seen, Tversky and Kahneman (1986, 1987) would have us draw this conclusion, which, in turn, has prompted a barrage of philosophical responses more virulent than Cohen's, trying to undermine the significance of Tversky and Kahneman's experiments (cf. Kyburg 1987, Levi 1987). The curious feature about this exchange is that all the parties make an unnecessary assumption about what it means to "naturalize" epistemology, namely, to make it coextensive with empirical psychology, whose goal it is to model the cognitive processes of the average individual knower (Quine 1969). Thus, the psychologists argue that naturalizing epistemology eliminates rationality, whereas the philosophers argue that the grounds of rationality cannot be naturalistic ones (cf. Putnam 1983, ch. 13). The result is the epistemological round of the debate between "is" and "ought," a posteriori and a priori: who has got the facts versus who is committing the fallacy. However, this latest version of the debate takes on more alarming proportions because the psychologists claim that our actual cognitive processes are not merely suboptimal by some preferred normative standard but are downright dysfunctional.

Yet, at the same time, this debate leaves out most of the real normative naturalism that preoccupies philosophers of science today, which we examined at the beginning of this chapter. Whether it is Laudan, Giere, or Goldman, most normative naturalists use the facts (historical or psychological) as a background constraint within which their theories of knowledge and rationality are then proposed. Indeed, philosophical theories of scientific methodology have generally been animated by pragmatist considerations, however abstractly expressed. In other words, certain theories of rationality are preferred over others in virtue of their probable real world consequences for knowledge production (cf. Nickles 1987). Laudan (1987) was quite explicit in casting the search for such a theory as an inquiry into the most efficient means for pursuing some desirable outcomes, whether past theory choices or ones in the future. However, neither Laudan nor his opponents have been particularly forthcoming on the sorts of entities that are supposed to instantiate the method: individual scientists? scientific communities? a third-person observer such as the historian of science?

Ironically, in spite of his generally negative views about the sociology of knowledge, Laudan (1977, 1984) has been a consistent proponent of the view that most philosophers of science silently believe, namely, that the *scientific community* is rational in deciding by a certain date to pursue one research tradition over its competitors. And thus, without doubting that Newton was a great scientist, Laudan et al. implicitly concede that Newton's *own* decision to establish an alternative tradition to the existing scholastic and Cartesian ones in natural philosophy does not *automatically* recommend itself as a rational move (cf. Baigrie 1988). But for all its surface irony, the link between consequentialism and sociologism in the search for knowledge makes perfect sense. After all, it is the *philosopher* who adopts a consequentialist stance to methodology by examining the long-term trajectory of the history of science; if she sees that desirable epistemic outcomes are produced, she then recommends that individual scientists adopt the methodology as a matter of principle, that is, they should not waver when things do not work out in the short term. In effect, philosophers are supposed to legislate the principles that scientists use as the basis of their research adjudications (cf. Rawls 1955). This is why a sociology of knowledge wedded to the philosopher's enterprise, such as Weber (1954) and Merton (1968), will to tend portray science as a Kantian moral community.

Philosophers have not helped clear up the ambiguities here by remaining silent on what may be called "ontological" questions about knowledge and reason. As a result, they make it seem as though there is no real problem of what knowledge *is*, only a question of whether we can get any of it, which means whether we can justify claims to knowledge (cf. Pollock 1986). However, if acquiring knowledge does not lead philosophers to think of knowledge as a substance, *then maybe we should think instead about trying to get rid of some*. Consider these two questions: (1) Can we prevent ourselves from getting the sort of knowledge that we will later want to get rid of? (2) Can we get rid of knowledge once we have gotten some? Appearances to the contrary, these are not just fanciful ways of asking, respectively, whether we can avoid falsehoods altogether, and if not, whether particular falsehoods can be identified. Rather, my two questions presuppose that once produced, knowledge has an existence sufficiently independent of its producers that merely changing one's mind about the truth value of a proposition will not be enough to eliminate it from one's knowledge base. Memories will need to be erased, patterns of speech altered, habits broken, and books rewritten or written anew. If knowledge had no material component, then it should not matter whether we avoid falsehoods from the start (as in the Cartesian project of rational foundations) or eliminate them as we go along (as in the Popperian project of rational criticism). In either case, we should end up with the truth.

But clearly, there is a psychological preference for never incorporating falsehoods, as if once incorporated, the falsehoods might become unremovable. This would help explain Descartes's own epistemological anxieties as well as why Popper's dictum of "start anywhere, but falsify from there" has won relatively few adherents. In analyzing how policy preferences tend to be ordered, social choice theorists have noted a cognate phenomenon, *hysteresis*, the psychological tendency to believe that it is worse to lose than never to have gained at all (Hardin 1982, pp. 82–83). Epistemological hysteresis is most evident in the intuition that skepticism, with its total sense of cognitive resignation, is psychologically more tolerable than fallibilism (Naess 1970). The perceived asymmetry between gains and losses may be diagnosed in several ways, each pointing to the crucial role played by the material character of knowledge, especially the pain that results from the persistent memory of the proposition formerly held to be true, as well as an awareness of the waste in mental effort implied in rejecting a proposition that one had originally spent some time coming to accept. Under these circumstances, we should try to answer (1) in the affirmative so that we will not need to pose (2).

What (1) asks us to consider is whether it is possible to have a policy of "planned obsolescence" toward knowledge. Can we anticipate when a theory currently regarded as true will be shown false, when it will have been applied to a range of phenomena for which it cannot adequately account? Taking a cue from Thomas Nickles (1986), we may call this test for the theory's effective scope, "negative heuristic appraisal." In the physical sciences, it has long been observed that one consequence of a paradigm shift is that the successor theory delimits the effective scope of its predecessor. Thus, relativistic mechanics explains not only why Newtonian mechanics was as true as it was but also why it must be false beyond a certain range of phenomena. However, only the staunchest Whig historian of science would want to claim that scientists have generally *tried* to anticipate the bounds of their knowledge claims. If they did, then there should be evidence of research programs with predefined limits and a promise to close up shop once those limits were reached. By contrast, we find that scientists usually design research programs as boundless enterprises, which are brought to a gradual or abrupt halt through the *unanticipated* consequences of their own and their competitors' activities. But would trying to anticipate these consequences improve matters? Not necessarily, as the following potted history illustrates.

It was obvious to scientists in the Late Alexandrian, Medieval, and Renaissance periods (A.D. 200–1600) that Aristotle's cosmology and Ptolemy's astronomy were inconsistent. From a Whiggish perspective, we might be tempted to say that if the "medievals" (to give these disparate peoples a suitably pejorative label) really had wanted to get to the truth of the matter, they would have constructed something like crucial experiments and started to discard some parts of these theories as false and perhaps even entirely replace one or both of them. Although some changes were made to the systems of Aristotle and Ptolemy during this period, they were made for reasons unrelated to crucial experiments, but more in the interest of defining the domain of applicability for each discipline more carefully. Astronomy turned into a practical concern, while cosmology became purely metaphysical. After a few generations, people trained in one field were no longer expected to have expertise in the other. Among the consequences of these moves was the medievals being led to believe that reality was a many-splendored thing and that knowledge was increasing by leaps and bounds. Another consequence was the belief that it was possible to teach the conceptual structure of knowledge by teaching its historical development, even to the point of indicating where new domains of knowledge could and could not appear, since a new domain would have to stand in a some determinate relation to the domains already, and always to be, in place (hence, the medieval preoccupation with "trees" and "maps" of knowledge). Also, this dovetailing of the conceptual with the historical made for a fluent pedagogy (cf. Kuhn 1981 for a Piagetian twist to this story). For if we are interested in defining a theory's domain of applicability, the context of justification is unlikely to stray very far from the context of discovery (cf. Nickles 1985). Indeed, if this 1,400-year period has received the reputation of being intellectually stagnant, it is not because nothing was being *added* to the "Book of Knowledge" but rather because nothing was being *eliminated* from it. One of the interesting features of Galileo's Inquisition was that the Jesuits accused him of being a scourge to learning, and it is easy to see why. Galileo, never the pluralist, wanted a confrontation between certain theories over some issue and then to make unequivocal statements about their truth and falsehood, and finally to eliminate all the falsehoods. For all their pluralistic ways, the Jesuits could not have what Galileo suggested, which was the possibility of a genuine intellectual revolution, which would satisfy Galileo's urge *to burn a few books* (cf. Fuller & Gorman 1987).

What is the moral of this tale? It would seem that planning epistemic obsolescence is much easier said than done, especially when one of the things *not* done is to account for the material component of knowledge. For by the time Galileo was brought to trial, the Jesuits had cultivated a highly refined sense of hysteresis, which led them to doubt that new knowledge could ever eliminate old knowledge, and if there was a threat of elimination, then it could only mean that the new knowledge was not really knowledge at all. This potted history would therefore seem to suggest that we admit the unfeasibility of receiving an affirmative answer to (1) and proceed to (2) in good Popperian fashion.

But once we factor in the material component of knowledge in addressing (2), we again start getting into problems. In particular, when inquirers engage in a strategy of falsification or eliminative induction, *where* do the false knowledge claims go once they have been eliminated? As it stands, philosophers have been content with equating "eliminate" and "recognize as false," somehow assuming that once unmasked, a false claim disappears out of its own accord, never to return again. Indeed, this assumption is the flipside of the inertialist image of reason that propels the internal history of science. Whereas the truth naturally persists, falsehood must be artificially maintained; once the artifice has been revealed, falsehood loses its support system. This pretty picture has had its share of skeptics, though few are to be found among the ranks of philosophers. Rather, they are to be found among those who take the material character of knowledge seriously enough to realize that falsehoods do not instantly self-destruct. These are the brainwashers, bookburners, and operant conditioners of the world—often a morally unsavory bunch, but one with keen insight into the persistence of error (cf. Jansen 1988). At least, David Hume had learned something from them: He demanded that any book not containing mathematical or empirical claims be cast into the flames (Thiem 1979, Fuller & Gorman 1987).

If we think about these matters a little less drastically, epistemic elimination might be taken to mean that a false knowledge claim ought to be removed from the cutting edge of inquiry, but otherwise set aside for safekeeping, perhaps by a discipline in the humanities. Indeed, the fact that knowledge claims are periodically eliminated in this sense allows such humanistic disciplines as the history of science to have a subject matter with something more than purely fictional or imaginative content. However, this story is too good to be true, since were it true, we would then expect the humanities to preserve everything in our epistemic history, which they in fact do not, as illustrated by the many disputes over the formation of reading canons and the definition of "classics." Moreover, even if we confine ourselves to what the humanities actually preserve of our epistemic past, who is there to prevent these supposedly obsolete texts from making their way back to the forefront of inquiry at a later date and in an unwitting form? After all, a case could be made that scientists (e.g., Prigogine & Stengers 1984) who want to use their research to stage a comprehensive conceptual revolution would come up with more radical conclusions if models for opposing the status quo were not so readily available in the guise of defunct science. (Admittedly, this speculation is, in an important sense, unintelligible, since if all traces of defunct science were destroyed, it would be impossible to tell whether, say, Prigogine would reinvent Aristotle and Epicurus in the course of extrapolating from his own thermodynamic theories.) In the end, we might just have to admit that unless drastic measures are taken, knowledge simply cannot be eliminated, but merely shifted, with varying degrees of control, from place to place on the epistemic map (cf. Fuller 1987c).

9. Or Maybe Broken Down

In pursuing the sort of things that embody knowledge, we may have taken the wrong tack. Rather than thinking of the ontology of knowledge as a *matter theory*, maybe we should think of it as a *field theory*. In that case, knowledge is a property distributed over a region of space-time, not aggregated in determinate locations. One consequence of this move is to make societies, instead of books and other textual artifacts, the proper bearers of knowledge.

As heirs to Descartes, we have developed a philosophical reflex to look inside a person's mind (or brain) as the first move in the search for knowledge. However, the philosophically naive (e.g., introductory philosophy students) are just as likely to begin by turning to books, databanks, and skillful practices in their epistemic inquiries—and one cannot fault the naive here for failing to abide by the ordinary language use of "knowledge." A more sophisticated view of the matter, which nevertheless draws on some of this intuitive support, is that all interesting epistemic properties are relational (cf. Bechtel 1985). In other words, to think that "I have knowledge" entails that my knowledge is encoded somewhere in my brain may be to fall victim to an illusion of surface grammar. Instead, my having knowledge may involve an interpreter crediting me with a certain range of possible utterances and actions with which she expects, or at least would permit, me to follow up my current utterances and actions. If I fail to act within these expectations and permissions, then the attribution of knowledge is withdrawn as having been made by the interpreter in error.

Now, if we were to assume that all recognized members of a community deal with each other in this fashion, then, strictly speaking, the brain scientist would find *my* knowledge in the brains of those who credit me with knowledge, whereas my own brain would be encoded with the knowledge I credit to others. In that case, the epistemic use of "my" is as a term of license, not of ownership. (At most, one could say that I "own" the evidence—the actions and utterances—on the basis of which others credit me with knowledge.) An important implication of this view is that knowledge is not neatly parceled out into clearly defined individual bodies, but is rather a field property, whose distribution among the members of a community shifts as mutual interpretations shift, with the result looking very much like a symbolic interactionist's idea of a "collective mind" (cf. Mead 1934). To ease you into this new conception of knowledge, imagine the following steps by which we might systematically divest my knowledge from my body and distribute it among the members of my community, whose own identity is itself ultimately dispersed:

(1) I am in a privileged position to know what I mean and that puts me in a privileged position to know whether it is true.

(2) I am in a privileged position to know what I mean but in no such position to know whether it is true. That is for my audience to determine.

(3) I am in no privileged position to know either what I mean or whether what I say is true. My audience is in a privileged position to determine those things. But I am in a privileged position to determine who my audience is.

(4) I am in no privileged position to know what I mean, whether it is true, or even who is and is not part of my audience. In fact, the audience is in a privileged position to determine my own identity as a speaker.

(5) Not only am I in no privileged position to know what I mean, whether it is true, or who is and is not part of my audience, but also the audience itself is in no privileged position to determine who does and does not belong to it, which implies that my identity as speaker is at best fragmented.

Although the process unfolded here would be recognizable to such French post-structuralists as Foucault, Lacan, Derrida, and Lyotard as "de-centering," "dispersing," or "disseminating" the subject (cf. Soper 1986), analytic philosophers have also lately courted strikingly similar conclusions after realizing that the content of a thinker's thoughts is not exhaustively determined by what goes on in her mind (Woodfield

1982, Pettit & McDowell 1986). The sorts of considerations that have moved the analysts in a more superindividual direction are captured in Hilary Putnam's (1975, ch. 12) "Twin-earth" thought experiment. Putnam asks us to consider two substances, H_2O and XYZ, with exactly the same phenomenal and functional properties as water. The difference is that one is found on Earth and the other on Twin-earth, and hence each is made of the sort of stuff found on the respective planets. Putnam argues that when I speak of "water," whether I am talking about H_2O or XYZ will depend on the planet to which I am taken to be making implicit reference. As possibilities (1) through (5) illustrate, intuitions vary over how much epistemic authority I have, vis-à-vis my audience, in fixing this implicit reference. Literally speaking, the question of what I know boils down to who gets the last word on what I said (cf. Skinner 1957). Taken as five steps, (1) through (5) involve incorporating more of what would normally be called *context* into the *content* of my utterance. Analytic philosophers, following Putnam, use the expression "wide content" to capture this incorporated sense of content.

Now, to survey the range of possibilities briefly, we started with (1) the classic Cartesian position, which is also captured in Fregean approaches to semantics (e.g., logical positivism, but also—for self-determining thought-collectives—Kuhn and Feyerabend), whereby meaning fixes reference, and hence a change in meaning causes a change in reference; (2) Kripke and Putnam's "causal theory of reference" (Schwartz 1977), which detaches meaning from reference, thereby opening the door to various forms of scientific realism; (3) the late Wittgensteinian dependency on the original context of utterance for meaning, as pursued in recent years by Tyler Burge (1986); (4) the reconstitution and incorporation of context into ongoing traditions of reading, as developed by the Cambridge-trained political theorists Quentin Skinner, John Pocock, and John Dunn (e.g., Skinner 1969); (5) the complete reduction of both content and context to shifting aggregates of mutual evaluations (Lehrer & Wagner 1986) or interpretations (Davidson 1986).

Thought experiments aside, the problem raised by Putnam's Twin-earth example is central to any account of interpretation, especially when it comes to distributing the "burden of error" between the interpreter and the interpretee, a point that will be stressed in Chapter Four as crucial to making sense of scientists' behavior. Faced with some strange behavior on the part of the scientist, does the interpreter then conclude that she has misunderstood something that is rational on the scientist's terms, or rather, that she has spotted errors that have escaped the scientist's notice? In either case, the interpreter is trying to do something that the French post-structuralists believe cannot be done, namely, to find a seat of epistemic authority, or a level of analysis at which the principles of rationality work. If individuals taken in Cartesian isolation are as irrational as the psychology experiments suggest, then maybe—so more analytically inclined interpreters suppose—the unit of reason lies in a superindividual unit, subject to a specific delegation of epistemic authority among various individuals in various situations. This is the hope even of an analytic philosopher with as dispersed a view of epistemic authority as Keith Lehrer, who would decide knowledge claims by a weighted average of the evaluations that members of a community make of both the candidate claims and of each other's evaluative abilities. He nevertheless holds that, in the long run, a consensus will stabilize, as members continually adjust their evaluations to match those of the other members whose abilities they respect (Lehrer & Wagner 1986).

10. Or Maybe We Need to Resort to Metaphors: Everyone Else Has

Levi-Strauss and Piaget to the contrary, primitives and children are not alone in confusing signs and referents, words and things. Indeed, we scientific sophisticates may be guilty of an especially deep form of confusion that makes it difficult to track down which are the signs and which are the referents. Levy-Bruhl (1978) originally used the term "participation" to diagnose the roots of primitive animism, namely, the tendency to attribute properties of a natural referent to its conventional sign, as in the events described in a curse being destined to occur, simply in virtue of the curse being uttered. In the case of moderns, however, the problem is almost the reverse, namely, that we cannot prevent the properties of our verbal images from contaminating the things of which they are supposed to be images. In fact, examined from a long-term perspective, modern science can be easily shown to have developed no clear way of distinguishing literal from metaphorical truth. After considering a couple of potted historical examples, one centering on classical mechanics and the other on evolutionary theory, we will then move on to discuss the pregnant confusions currently centering on cognitive science.

It is a commonplace among contemporary social science methodologists to argue that late nineteenth century theories of collective behavior illicitly traded on an ambiguity between the well-founded quantifiable physical forces of Newtonian mechanics and a more occult but nonetheless suggestive concept of "social forces," whose "kinematics" and "dynamics" had been touted as early as Auguste Comte (cf. Sorokin 1928, ch. 1). Indeed, the effects of this ambiguity are felt even today, as social historians blithely assume that "mass movements" (a play on the physicist's matter-in-motion) are proper objects of inquiry. An especially common form of social history explains popular uprisings in terms of demographic pressures on scarce resources, which certainly brings to mind the account of temperature increase given by Boyle's Law. But before prematurely concluding that the sociologist's entities are mere reifications of the physicist's lexicon, we need only recall that Boyle's and Newton's idea that ours is a universe governed by laws relating active forces and passive masses was itself drawn from a leading seventeenth century monarchist model of political order (Jacobs 1976).

The history of twentieth century social theory and social science has been plagued by confusions over whether societies are supposed to evolve in a manner analogous to either Lamarckian or Darwinian accounts of biological evolution, or whether the evolution of societies actually constitutes the final stage of biological evolution. (Incidentally, the same confusion applies, mutatis mutandis, to evolutionary epistemology, especially the significance of Popper's dictum, "Theories die in our stead." Is theory testing merely analogous to organisms facing selection pressures, or are theories literal extensions of the human organism that contribute to its evolutionary fitness? [cf. Ruse 1986, ch. 2].) Because this ambiguity first arose at the turn of the century, it is sometimes portrayed as being like the confusions over "social forces," whereby a more developed science is misappropriated to model phenomena in a less developed science, and then passed off as a reduction of those phenomena. However, this diagnosis works only if one neglects Lamarck's and Darwin's own respective reliance on the leading French and British social theories of the Enlightenment: Lamarck on Count Condorcet's account of human progress through increased education and self-determination, and Darwin on Adam Ferguson's explanation of the emergence and maintenance of civil society in terms of the conventional reinforcement of initially chance events (Richards 1987, ch. 1). That Lamarckianism has had political associations with socialism and Darwinism with capitalism should then come as no surprise, given that Condorcet and Ferguson were the eighteenth century forerunners of exactly these ideologies (ch. 6).

Finally, the case of cognitive science is one where the object of inquiry has shifted right under our very noses. Sometime during the 1970s, the metaphorical and the literal—the modeler and the modeled—switched roles, with a consequent shift in the balance of power between the disciplines involved in cognitive scientific inquiry. Prior to this period, the computer was said to *simulate* thought, which occurred, in its most robust form, in human beings. Computer models of cognition were thus presented as more or less gross simplifications of our complex reasonings. Indeed, this was why it had made perfect sense to say that computer scientists such as Marvin Minsky and Allan Newell were engaged in "artificial" intelligence research (Gardner 1987, ch. 6). After this period, however, the computer is often said to have at least the "in principle" ability to *instantiate* thought in a purer form (i.e., as pure computation) than can be achieved by the excessively noise-filled medium of the human being. Our deep complexity has now become, in effect, a mechanical deficiency, which makes us a fertile ground for hypotheses about the nature of thought, but ultimately just a first approximation of what the computer will fully realize. In terms of the balance of disciplinary power, computer science originally worked in the service of experimental psychology, but nowadays the experimental psychologists are subservient to the computer scientists.

The one person who embodies this shift most closely in his own career trajectory is the Canadian Zenon Pylyshyn who is, not surprisingly, among the most metatheoretically astute people in cognitive science today. Pylyshyn, who received an M.Sc. in computer science and a Ph.D. in psychology (both from the University of Saskatchewan in the 1960s), began his career conducting research on the application of computers to the analysis of psychiatric interviews, which is to say, he started by treating computers and humans as separate albeit interactive entities. A quote from this early period conveys the sense in which he still regarded computers merely as artificially intelligent creatures trying to catch up with humans:

> One may be interested in trying to make a machine do "intelligent" things for at least two reasons. One is to see how far one can push the limits of intelligent behavior in a machine. There is still amazement and a sense of achievement for every problem hitherto in the exclusive province of human competence which we can make a machine solve. The second reason is to learn something about human thought itself—to try to understand human cognitive functioning by constructing models of this function and then attempting to simulate these on a computer. (Pylyshyn 1970, p. 219)

Over the years, however, Pylyshyn has reconceived the locus of his research, so that nowadays he freely speaks of humans instantiating the sorts of rule-governed procedures that one would normally program on a computer. Indeed, his computational model of imagery is counted (by a leading journal in the field) as a piece of research in *psychology*, even though it treats the human experience of thinking in images as mere epiphenomena on the underlying processes of symbol manipulation (Pylyshyn 1973). All this has led Pylyshyn (1984) to argue that cognitive science aims to uncover the laws governing the natural kind *cognizer*, instances of which equally include digital computers and human beings—*not* humans primarily and then computers in some derived, metaphorical sense. Moreover, the computer may offer the purer case, especially if we conceive of the relevant laws as computable functions, precedent for which can be found in what Hobbes, Leibnitz, and Boole, among others, called "the laws of thought." And like his philosophical predecessors, Pylyshyn understands these laws as being *both* normative and empirical—this in spite of the Humean vexations over the sorts of inferences that can be licensed between "is" and "ought." But as we shall eventually see, Pylyshyn's search for the essence of thought gives today's cognitive scientist a distinctive slant on this problem.

One classical way of taking the claim that the laws of thought are at once normative and empirical is in terms of the competence–performance distinction, whose latter-day revival is due to Noam Chomsky. On this view, our actual cognitive performance is typically only an imperfect representation of our innate competence, the full extent of which can be elicited by the right verbal protocols (e.g., intuitions about the grammaticality of a sentence) and, of course, training (e.g., in logic, grammar). Crucial for this view is that the empirical cues appear impoverished in relation to the response produced in subjects, which would strongly suggest that competence is indeed being elicited rather than performance merely being enhanced. Although our ability to intuit the grammaticality of sentences we have never previously heard is suggestive of a Meno-like rational unconscious, the real evidence for our being more competent than our performance indicates comes from the *speed* with which our performance improves upon learning even a little grammar.

However, when we turn from grammar to canons of reasoning—ranging from universal deductive and inductive principles to more domain-specific ones (including self-interested principles of rationality, such as utility maximization)—we find that performance is much less tractable (e.g., Elster 1986). As was noted in earlier sections, experts do almost as poorly as novices in a wide variety of reasoning tasks, and training prior to the task improves performance only under highly specific circumstances (cf. Arkes & Hammond 1986). These results have made many leading experimentalists in cognitive and social psychology doubt that we possess the competence to empirically realize *any* normative account of rationality—in which case, why continue proposing normative theories?

11. Could Reason Be Modeled on a Society Modeled on a Computer?

It is worth repeating that the main protagonists of this debate, Amos Tversky (for the skeptical psychologists) and L. Jonathan Cohen (for the believers in rationality), presume that they are fighting over the cognitive status of the solitary human being. We have seen that this has led to several interesting turns, which have served to cast aspersions on the individual as the unit of rationality. Instead of despairing of the possibility of a normative psychology, why not simply say that the isolated individual is not its proper object of study? A sociological approach to rationality would seem to be the most obvious beneficiary of such a move, as illustrated by the attempts of a clinical psychologist, David Faust (1984), and a philosopher of mind, Stephen Stich (1985), to transcend the individual in their search for higher normative ground.

Taking his cue from Popper, Faust argues that, as a community of cognitively limited creatures, we can make the most of our biases by being our own conjecturer and our neighbor's refuter. The net effect of the interaction should be the survival of the epistemically fittest ideas. This approach recalls Durkheim's (1933) mechanical solidarity, in which each individual is an imperfect microcosm of the emergent social order, contributing in her own limited way to each of the knowledge production tasks. By contrast, Stich's conception of social rationality resembles the division of labor associated with Durkheim's organic solidarity, in that each individual is now made to excel at a different task, perhaps with some overlap in assignments to facilitate communication. In short, we are experts in our own conceptual region but must defer to others when inquiring beyond that region. Precedent for Stich's view may be found in Donald Campbell's (1969) "fish-scale model of omniscience," which was originally proposed to foster individual innovation (i.e., to explore the conceptual space left untouched by two neighboring specialties) and overall epistemic coverage, not merely disciplinary channeling for a community of subrational minds.

The spirits of Faust's and Stich's proposals are as far apart as the difference between critics and experts as exemplars of knowledge production: Faust stresses quality control (i.e., that the process is rational), whereas Stich is more concerned with maximizing productivity (i.e., that knowledge is actually produced). Despite these differences, however, Faust and Stich agree that individual human beings are not the exclusive bearers of cognitive properties. Both portray rationality as an abstract characterization of the structure of human interaction in the course of knowledge production (i.e., as a type of Durkheimian solidarity), whereas knowledge itself is simply the emergent product of this interaction. In that case, to return to the original problem of relating empirical and normative senses of the laws of thought, an individual's actual cognitive performance is to be evaluated in terms borrowed from structural-functionalist sociology, that is, by the contribution that the performance makes to the overall maintenance of the knowledge system. Metaphysically speaking, the individual is thus judged as a part in relation to the whole—not as a particular in relation to a universal (which would have made her out to be an imperfect instantiation of the essence of rationality [cf. Ruben 1986]). It would seem, then, that the sociologisms of Faust and Stich provide an alternative to Pylyshyn's move to treat the cognizer as a universal of which humans and computers are particular instances.

However, the most fundamental epistemological questions remain unaddressed by the Faust–Stich strategy. Assuming that knowledge is indeed a product that emerges from the interaction of individuals, where should we look in the knowledge system to find out how much and which kind of knowledge has been produced at a given moment? In computer terms, where is the epistemic community's "output display?" Given the cognitive limitations of each member of the community, there is no reason to think that sampling any of their opinions is likely to give an adequate answer to this question. Indeed, there may be a great deal of disparity in what individuals believe has and has not been epistemically established. According to Latour (1987a, ch. 6), the most heated power struggles in knowledge production concern just this issue, whereby groups compete to become canonical output displays, or "centres of calculation." Moreover, an individual's expressed beliefs may contradict her actual (e.g., citation) practices. If there is this much indeterminacy about what the state of knowledge is in a community at any given point, then it follows that the relative merits of various procedures for producing knowledge will be equally indeterminate. Of course, this entire problem could be treated as a technocratic issue, to be solved by setting up a bureau that surveys the knowledge production process, integrates its findings, and periodically broadcasts them from a central location for all the knowledge producers to see. The issue, then, would be the amount of resources that would have to be reallocated to this project from ordinary knowledge production—and whether it would be worth the cost.

In a limited way, the technocratic solution has already been undertaken by the Institute for Scientific Information in Philadelphia, which publishes the *Science Citation Index* and has recently embarked on publishing an *Atlas of Science*, whereby the narrative structure implicit in the citation patterns of the top twenty-five or so articles in a discipline is used to define the discipline's research trajectory (Small 1986). By "implicit narrative structure," I mean the convergence that emerges across these articles as to which pieces of research "explain," "support," "refute," or "supersede" other pieces (cf. Cozzens 1985, for disciplinary differences in the amount of convergence). That it is possible to discern such a narrative structure among the articles reveals the extent to which each author contributes to the growth of knowledge in two ways: not merely by adding another fact, but by treating the new fact (much as a Gestalt psychologist would) as an opportunity to reconfigure the state of knowledge in the field. This dual function is most evident in the literature reviews that introduce most articles in science.

Contrary to the folk wisdom concerning citation practices, a literature review is typically not the author's intellectual autobiography of what she read before writing the article that follows. Thus, the literature review should be taken as part of, not the context of discovery, but the context of justification—or, in Thomas Nickles' (1985) more precise sense, *discoverability*. That is, the review lays out the work in the field that provides what the author regards as a canonical framework within which her own work is the natural outcome. The point of the review is to make the author's work invaluable to other researchers by portraying it as having resulted from a strategic combination of important research. Since few, if any, of the author's readers can check on her actual research practices, they are forced to evaluate the article almost exclusively on the plausibility of her configuration of the field. If we imagine that everyone engages in this practice of providing a canonical narrative for her field, then all of the following consequences are likely:

(i) The sociologist cannot predict what the members of a discipline have read, simply on the basis of what they cite. On the contrary, there is reason to believe that highly cited sources, ones that any reader would expect to find in the literature reviews of a particular research area, are among the least consulted works. They find a place in the reviews mainly because of their known strategic value.

(j) Nevertheless, it is still true that the highly cited sources—and their authors—are officially the most highly prized in a field. In addition, regardless of the private lives and opinions of these authors, historians and newcomers have the network of citations in the literature as their primary, and often exclusive, exposure to the field's body of knowledge.

(k) Moreover, would the members of a discipline *want* to change the schizoid situation described in (i) and (j)? The need for members to make strategic use of each other's research in order to gain attention to their own ensures a certain amount of constraint on how old knowledge is appropriated in producing new knowledge, which, in turn, contributes to relatively consistent displays of the state of knowledge in the discipline at any given moment, as in the case of uniformly written literature reviews.

(l) What all this shows is that if the sociologist confines her inquiries to the citation practices of a scientific community, she may well arrive at a canonical output display. Ironically, the output display would have the detachment from local contexts of research that characterizes objective knowledge. But such a display would still probably not be a good predictor of what the practitioners themselves would say about what has and has not been epistemically established at a given moment. In a sense, then, the output display defined by a discipline's citation practices constitutes knowledge more for those outside the discipline than for those within.

It would seem that, in the end of our sociological analysis, Pylyshyn has been indirectly vindicated. In comparison with a computer that displays its output on demand, a society turns out to be an imperfect instantiation of a cognizer, since only by a massive redistribution of effort can a society determine its knowledge state at a given moment. And even confining ourselves to the narrative structure of citation practices, we find that it is best seen as the deliberate and ongoing rational reconstruction of a discipline's research trajectory, admitting once again that a discipline, taken merely as the aggregate of its practitioners' opinions, does not naturally gravitate toward a determinate state of knowledge.

12. Could Computers Be the Very Stuff of Which Reason Is Made?

If approaching Pylyshyn's cognitive essence is so removed from both the psychological and the sociological capability of human beings, then we might expect Pylyshyn's work to suggest a new sense in which the laws of thought are both normative and empirical. Indeed, I submit that the novelty here is that Pylyshyn suggests turning the idea of computer as metaphor exactly on its head: that is, no mere model, the digital computer simply *is* the empirical realization of the laws of thought, in terms of which human cognition is then evaluated and held accountable. If the computer is still said to be a "model," it is more as an archetype. How is that possible? The Argument For Computationalism (AFC) goes as follows:

(m) That there exist laws of thought, or canons of rationality, is demonstrated by our ability to recognize (i.e., discover) that certain principles, like Bayes Theorem in hypothesis testing and utility maximization in economics, promote more desirable outcomes in their respective domains—at least in the long term—than other possible principles.

(n) That human beings are not particularly good at obeying the laws of thought is also evident. It is significant, then, that we discovered these laws, not by comparatively testing the empirical consequences of following through on candidate principles, but rather by the sort of thought experiments familiar to positivist philosophers of science and neoclassical economists, experiments that do not submit the inquirer to any "real world" constraints.

(o) Assuming that the laws of thought are indeed intelligible (as [m] suggests), if we can recognize them through roughly aprioristic means (as [n] suggests), yet we also seem to be constitutionally ill-disposed to carrying out their procedures, then we have a good reason for thinking that the laws of thought have an existence independent of our own (cf. Peacocke 1988).

(p) Therefore, if rationality is at all empirically realizable, we should expect that there is something else in the world that is, strictly speaking, more rational than we. That thing (or class of things) would seem to be the computer.

(q) As a result, when we try to make sense of ourselves as rational, in both folk and scientific psychology, we are trying to explain as much of our behavior as possible in computational terms. In effect, the yardstick for measuring our psychological knowledge is its computer programmability. Thus, it is no accident that the features of our psyche that most strongly resist computational analysis, such as emotions and moods, are the ones that psychologists generally see as being least understood or rational.

AFC has several distinctive features, three of which will be considered. First, for all its initially Platonic overtones, AFC's thrust is really quite consistent with the Marxist themes of humanity as Homo faber and alienation as our natural mode of self-understanding. For what is being claimed here is that after a certain form of technology ("computers," which is to say, any computing device) achieved independence from its human construction, it became the standard against which we remade ourselves so as to more closely approximate it. This process of "remaking" is none other than the very process of self-understanding. To put the point bluntly, before primitive computers (e.g., the abacus), we had no canonical means of attributing rationality to ourselves and hence could not render ourselves proper objects of inquiry. The only way to give a Platonic gloss on this point would be to suppose that the first person to construct a primitive computer had a fully formed idea of computation, rather than (what is more likely the case) simply a vague notion that was designed to solve some specific counting problems, which only subsequently became the standard against which we measured our own rationality (cf. Fuller 1988b, ch. 2).

Even the appeal to a priori knowledge that seems to be made in steps (n) and (o) of the argument can be read in the sort of technological terms that would please the Marxist. Suppose a philosopher has the a priori intuition that consistently applying Bayes Theorem would produce a better long-term track record of hypothesis evaluation than any other scientific methodology. This intuition should be analyzed as a prediction of the output that would be displayed by a computer that has yet to be constructed. This machine would have a wide range of competing methodologies operate on a wide range of data from the history of science, retrospective science policy (i.e., what a Lakatosian would have done), and current science policy. The intuition is the prediction that the Bayesian would come out ahead in the end. Notice that if this is a correct analysis of the situation, then a priori knowledge is distinguished from the ordinary a posteriori variety by being a second-order prediction, i.e., a prediction (of the computer's result) about a prediction (that the computer will be built). By contrast, a posteriori knowledge normally involves only first-order predictions. In that case, we can subject the mysteries of a priori knowledge to the techniques by which we analyze and improve our empirical predictive capabilities. Indeed, we can do all this within a scientific realist epistemology, though one divested of any trace of Platonism.

The second noteworthy feature of AFC is that it tells against philosophers who have been skeptical about the possibility of a cognitive science (e.g., Taylor 1985, Dreyfus & Dreyfus 1986, Searle 1984.) Typically, these philosophers portray cognitive science as trying to *reproduce* human thought processes on a computer, as if there already existed ways of making sense of those processes that did not make use of computational models (even if only the rough syntactic sketches in which hypothetico-deductive explanations are normally couched in science). But rather than providing an alternative account of such computationally intractable traits as emotions and moods, the philosophers in question appeal to phenomenology, which methodologically forecloses any inquiries into mechanisms that may regularly produce these traits, precisely by defining emotions, moods, and so forth as essentially involving the first-person experience of having them. Thus, even if psychology discovered the full range of behavioral conditions and physiological consequences in which our emotional lives transpire, that still would not be enough for the phenomenologist because it would not capture the essential subjectivity of having an emotion: i.e., what it feels like for a human to have emotion, as opposed to what it feels like for a computer to have emotion (cf. Nagel 1979).

At this stage of the inquiry, the devotee of a naturalized human science may raise two queries:

(r) Assuming that an irreducible difference would always remain between human and computer subjectivity, why must we suppose that it would be a conceptually interesting one? That is, would this phenomenological difference predict behavioral consequences that would significantly distinguish humans from computers?

(s) To what extent do the phenomenologists exaggerate the difference between human and computer subjectivity by confusing the *consciousness* and the *self-consciousness* of the human being in thought? Phenomenological accounts of human thought are so rich because the phenomenologist records more than just her experience; she records her experience of having the experience, which introduces a second-order of consciousness. Implicit appeals to this second-order are made when the phenomenologist says that a line of reasoning is not merely a string of syntactic tokens, but one that is semantically rich, i.e., it has meaning for the thinker. As the maverick cognitive scientist Eric Dietrich (1988, 1990) has argued, against both anti-computationalists like Searle (1984) and pro-computationalists like Fodor (1981), a computer can also be similarly programmed to interpret its first-order syntactic manipulations. But normally, when humans are thinking, they are just like the ordinary computer in running through a program without consulting an "internal manual" of what the steps of the program mean. (Here the burgeoning psychological discipline of "protocol analysis" provides ways of preventing this collapse of levels that comes from the uncontrolled use of phenomenology [Ericsson & Simon 1984]).

Once the metaphysical ante of this point is raised, the distinction between persons and nonpersons starts dissolving into the mists of social conventions (cf. Fields 1987, Woolgar 1985).

However, nowadays it is difficult to discuss these subversive issues seriously, largely because the phenomenologists are generally seen as exercising epistemic authority over our intractable psychic traits and thus are able to keep the terms of the debate systematically ambiguous, which in turn keeps the key concepts contested, and hence resistant to any canonical formulation that might lend itself to computational analysis. It is clear, especially from Charles Taylor's (1985) writings, that a particular political agenda underwrites the phenomenological project, namely, to create as much conceptual distance as possible between humans and non-humans, especially machines, in order to continue the nineteenth century Romantic Revolt Against the Routinization of the Spirit by Technique. Admittedly, it would be fair to guess that a completely successful computational analysis of, say, the emotions would not leave the emotions looking very "emotional." In fact, I imagine that it would look like a cognitive account of conflict resolution (cf. De Sousa 1987). Nevertheless, the fact remains that for all its intuitive pull, the phenomenological project has the ironic consequence of turning the human essence into a ghetto for precisely those features of ourselves of which we have the least articulable knowledge and, hence, of which we are likely to be most ignorant!

The word "articulable" turns out to be operative here, since the vanguard of the phenomenological movement, in the person of Hubert Dreyfus, has been devoted in recent years to showing the resistance of cognition itself to computational analysis. (The popular version is Dreyfus & Dreyfus 1986.) Dreyfus bases much of his argument on the "tacit dimension" of knowledge, that is, the fact that we seem to know much more than we can tell, even in principle. For example, we are immediately able to make a sense of the contents of a room that we immediately able to make a sense of the contents of a room that we have never previously entered, even though the background knowledge that makes this possible seems inexhaustible, when regarded as a list of articulated propositions—and especially when these propositions are presented as semantically networked with other related propositions that happen to be irrelevant in the given context. How do we access just the relevant ones so quickly? According to Dreyfus, this inarticulate knowledge is supposed to reflect our organic attunement with the environment, or in Heidegger's terms, "Being-in-the-world." Explicating this intuitive sense of situatedness in computational terms is often called *the frame problem* (Dennett 1987b), and to the erstwhile supporters of cognitive science (e.g., Haugeland 1984, ch. 6), it would appear to be the field's Achilles' heel.

The quick and dirty way of subverting Dreyfusian thinking that is suggested by the above considerations is to wonder whether our minds are *too deep* or *too shallow* to conform to the computer model. Indeed, an examination of the evidence from both experimental social psychology and ethnomethodology implies that we are not as epistemically deep as we would like to think. On the one hand, being equipped with a healthy sense of superstition, a high tolerance for error, and a faulty memory, people are uncannily good at "telling more than we can know" (Nisbett & Wilson 1977). On the other hand, when asked to articulate the taken-for-granted features of their ordinary discourse, people do not dismiss the question (à la Dreyfus) as inappropriate; rather, they become anxious—as if suddenly realizing that the ungroundedness of their utterances had been revealed for the first time (Garfinkel 1963). Such an aversive response would not have been triggered had the people not been convinced, prior to being asked, that they could have addressed the request for background information.

Anthony Giddens (1984) offers this analysis of the situation: "Why did Garfinkel's 'experiments with trust' stimulate such a very strong reaction of anxiety on the part of those involved, seemingly out of all proportion to the trivial nature of the circumstances of their origin? Because, I think, the apparently minor conventions of daily social life are of essential significance in curbing the source of unconscious tension that would otherwise preoccupy most of our waking lives" (pp. xxiii–xxiv). For our purposes, the relevant conclusion to draw from Giddens's remarks is that, *contra* Dreyfus, human beings are only superficially adapted to their environments. Instead of being deeply attuned to ourselves and our ambient surroundings, we "get by," it would seem, by dint of presupposing the competence of ourselves and others to such a large extent that we are

rarely moved to make the telltale inquiries that would quickly unmask
our deep ignorance and catapult us into existential angst (cf. Edwards
& von Winterfeldt 1986, on cognitive illusions).

The third and final interesting feature of AFC is its potential for
radicalization. As it stands, the argument equivocates on the status of
human beings as cognizers. Here are two possible accounts of our status:

(t) Human beings are computationally just as good as digital
computers. The point of scientific psychology, then, is to figure
out exactly what sort of computer we are. The phenomenological evidence suggesting that we do not fully partake of the
computer's essence simply defines the domain that is in need of
further empirical investigation. But assuming we have a fair
chance to experiment on these phenomena, they too should
ultimately turn out to be explicable in computational terms.

(u) Human beings are computationally not as good as digital
computers. The point of scientific psychology, then, is to
define the limits of our computational abilities so as to allow
more reliable computers to make up the difference. This leaves
the core of cognitive science to determine the epistemology of
the pure form of cognizer, the computer android. As for the
phenomenological counterevidence, Dreyfus et al. may be
right about its computational intractability, but only because
they have identified the ineliminable "noise," *not* something
deep, in our cognitive makeup (cf. McDermott 1987).

Pylyshyn's own position is clearly (t), which motivates his elevation of
computer modeling from metaphor to archetype. As Pylyshyn (1979)
sees it, computation is to cognition as geometry is to mechanics, that is,
in both cases, the former analogue provides the literal structure of the
latter. From a historical standpoint, this is a very apt comparison, since,
like many who today resist computational analyses of cognition, the
Aristotelians were notorious for treating mathematics as a useful way of
representing the physical world for certain purposes, but by no means
the basis of a systematic account of its essence (Gaukroger 1975, ch. 1).
Thus, Platonists and Pythagoreans were typically regarded as having
reified the metaphor of number, much as computationally oriented
cognitive scientists are regarded as doing today. Nowadays, of course, a
constraint on the very intelligibility of physical theories is that they be
mathematizable.

Nevertheless, in the original exchange that prompted the above
considerations, Pylyshyn noticed an Aristotelian response to the
"metaphor of information-processing" even by an avowed cognitive

science sympathizer, the scientific realist Richard Boyd (1979). Pylyshyn observed that what separated his view of the computer metaphor from Boyd's was the sort of "imprecision" that each supposed was appropriate to the scientific use of metaphor. Much like Aristotle, Boyd assumed that natural kinds—the sorts of things that may be modeled in a metaphor—are *substances*, although it would seem that, by definition, the two things compared in a metaphor are different precisely in substance. Thus, Boyd set himself up to see the computational metaphor as having inherently limited value, since a computer and a human being are indeed substantially different things, one a machine and the other an organism. In response, Pylyshyn argued that Boyd's focus on substance led him to overestimate the scientific significance of the surface differences between machines and organisms, when in fact the two may be isomorphic at some deeper level of *structure*. The natural kind cognizer is thus proposed as this deeper structure. Pylyshyn's more Platonic realism shifts the epistemic interest in metaphor from one thing's resembling another to the two things' sharing a common identity. However, he presents this as an empirical research program for the study of human beings. The aim is to specify the "functional architecture," which is, in effect, the programming language in which all human computations occur. The basic strategy is to find behavior that is resistant to change in background knowledge (i.e., "cognitively impenetrable"), since that would be a good indicator of the behavior's innateness and hence a basis on which to infer the programming language responsible for it (Pylyshyn 1984).

13. Yes, but There's Still Plenty of Room for People!

As an empirical research program into the nature of human thought, computationalism may turn out to be false. In any case, it is likely to be locked in a long and messy battle with the phenomenologists. But according to the (u)-interpretation above, this is to miss the truly radical import of AFC, which follows through on the conclusion that the standard of rationality literally lies outside ourselves and in computers, the ultimate cognizers. In that case, the anthropomorphic constraints of empirical psychology would seem to place an unnecessary condition on getting at the nature of these cognizers. What we need to practice, then, is what Clark Glymour (1987) has dubbed *android epistemology*. Let us recall the obstacle posed by the frame problem: How one solves this problem for computers may be completely different from, and unilluminating for, how one solves it for human beings. Indeed, in an exasperated exchange with Jerry Fodor, computer scientist Patrick Hayes (1987) recently observed that the computer version of the frame problem was originally conceived as a matter of programming the computer to do what it is supposed to do at the right times and places. That computers manage to perform as designed implies that the frame problem is, in practice, solved. But whether there are any general procedures for solving the frame problem beyond the engineering requirements of particular computer environments and whether such procedures are likely to reveal anything interesting about how humans adapt relevantly to their own changing environments (cf. Janlert 1987) are entirely separate and difficult questions, the solutions to which should not forestall the development of android epistemology.

It may seem from my last remark that efforts at developing android epistemology have so far been mired in debates with the phenomenologists. But such a conclusion would be misleading, since an android orientation is betrayed every time a philosopher attempts to divest human beings—be they ordinary or expert—of their cognitive authority. Often these divestment moves are seen as part of the latest wave of philosophical imperialism, but it is unclear just how much philosophers themselves stand to gain by them. A telling case in point concerns a pair of inductive logicians, Cohen (1986) and Kyburg (1983), who want to mitigate the pessimism toward normative theories of rationality that has followed in the wake of Tversky and Kahneman's experimental findings. Cohen argues that in cases where subjects are not patently being deceived by the experimenter, their responses, which Tverksy & Kahneman (1987) present as contradicting all the leading models of rationality, in fact conform to a type of inductive reasoning commonly used in juridical contexts, which philosophers have failed to explore because they have relied on natural science rather than law as the paradigmatic setting for induction. Notice that, in making his case, Cohen presumes that the intuitions of the experimental subjects are largely rational, that is, reliably tapping into some deep-seated competence for rational thought. What Cohen disputes is whether most psychologists and philosophers (not to mention the subjects themselves) have gotten a handle on the theory of rationality that best explains those intuitions.

In response, Kyburg questions why any epistemic authority should be conferred on even the subjects' responses. After all, just as subjects may have learned false folk theories to justify their intuitions, the intuitions themselves may be nothing more than routinized verbal responses, reflecting more on the contexts of conditioning and reinforcement than on any innate rational capacity. In effect, Kyburg is arguing that the malleability of human "rational" behavior is prima facie grounds for believing that human beings are not the standard of rationality (compare Pylyshyn's criterion of cognitive penetrability, discussed above). Here, then, is the implicit call for android epistemology. But philosophers benefit from this call only insofar as they can discover the laws of thought. Kyburg is reflexive enough to know that philosophers are no better than other mortals in their abilities to instantiate the laws in their own practice. It should come as no surprise, then, that he has increasingly taken to publishing in computer science journals. In short, when it comes to the study of androids, what the phenomenologist taketh away, the inductive logician giveth right back!

But what would an android epistemology look like—and in particular, how would it compare to cognitive science, as commonly understood today? The key to answering this question lies in the policy implications of android epistemology. To summarize the above discussion, what do we make of our inability to think like a digital computer? We have so far surveyed three basic options:

(v) *The phenomenological option:* Drop the computer as the model of human thought (e.g., Dreyfus).

(w) *The rationalist option:* Develop a pedagogy that will enable people's cognitive performance to realize their innate competence (e.g., Cohen).

(x) *The empirical option:* Treat the computer as a framework for conducting empirical research to see how much of human cognition can ultimately be explained by it (e.g., Pylyshyn).

Android epistemology now presents a fourth option:

(y) *The division of labor option:* Let computers do what they do best, and let humans do what they do best.

Option (y) is radically different from the other options in its denial that the computer and the human model each other's activities. In fact, (y) implies the very opposite, that is, that the computer and the human *complement* one another. This gives the research program of android epistemology a very distinctive trajectory. For in accordance with AFC, the android epistemologist is bound to discover "laws of thought," that is, rules for systematically producing epistemically desirable results in a given domain. But rather than trying to force these laws upon the human intellect (e.g., via statistics courses, aptitude tests, and ad hoc psychological hypotheses), the android epistemologist simply concludes that the laws govern the android mind of the computer, to whose authority the human defers when she is interested in some consequence of the laws. In other words, we are to treat the computer android much as we treat such instruments of perceptual enhancement as microscopes and telescopes. For example, although the ability to construct a telescope presupposes at least some rudimentary understanding of how the human eye works, that fact alone has never been sufficient to support a general policy of training the eye to approximate more closely the power of a telescope. Likewise, the android epistemologist would claim, the cognitive processes embodied in the ability to program a computer that can test hypotheses, calculate utilities, or (dare I say) do sums need not be reproduced in ourselves.

A very instructive history of science could be written about how human beings have learned to budget their mental resources more efficiently by gradually delegating to machines, as "prosthetic reasoners," rule-governed tasks that humans had previously performed only with great uncertainty and under much cognitive duress. Indeed, as humans have been able to construct each such machine on a reliable basis, the task performed by the machine has been rendered, in Pylyshyn's terms, "cognitively impenetrable." Thus, whereas the ancients, when relying on their own eyes, would argue incessantly about the nature of the objects of distant vision, we nowadays simply trust that the telescope registers data about distant things. Of course, optics is still a going concern in the physical sciences, but not because we think that its unresolved foundations are jeopardizing the epistemic status of visually obtained information. As Bruno Latour (1987a, pp. 2–3) would put it, the telescope's optics has been put in a "black box." And it is here that the cognitive impenetrability of the telescope's data makes a difference. For if you propose an astronomical theory today that is true only if the function of telescopes is radically reinterpreted, then you run a great risk of not being recognized as doing astronomy. Of course, telescopes occasionally break down, but the locus of these problems is taken to reside in the particular instrument, not in a host of deep metaphysical issues about the possibility of visual knowledge.

If we regard this scenario as generally representative of the history of science, then we should expect that a time will come when testing rival hypotheses is as cognitively impenetrable as doing one's taxes on a pocket calculator: The scientist would simply plug the relevant information into her hypothesis testing machine and *trust* that the hypothesis on the machine's output display is the one to pursue (cf. Meehl 1984).

By delegating more and more cognitive authority to computer androids, we are afforded greater opportunities to contemplate whether there is anything distinctively human about our cognitive situation. Indeed, this feature of android epistemology should please even phenomenologists, since it implies that our unique properties (if there turn out to be such things) are defined as non-computational, that is, the modes of thought in which computers cannot outperform us. Needless to say, what those modes of thought might be is subject to constant revision. For example, should a computer program be developed with the ability to generate hypotheses that current scientists regularly find worth pursuing in their research (i.e., the program does not merely generate proven winners in the history of science; cf. Langley et al. 1987), then, rather than denying the reasoning powers of the computer (as phenomenologists might be inclined to do), the android epistemologist would simply conclude that this is yet another cognitive burden that humans are not especially well-designed to bear (cf. Simon 1981).

I do not imagine that the cognitive burdens of doing science will someday be completely relieved by one mega-program, but rather they will be whittled down over time by a collection of expert systems, each of which will be able to outperform humans in generating pursuitworthy hypotheses in specific domains of inquiry. (Think of this by analogy with the Turing Test, so that when faced with just the two alternative hypotheses for a given set of data, practitioners of a given discipline prefer the hypothesis generated by the computer to the one generated by their colleague.) Moreover, as these androids divide and conquer the labor of producing knowledge, the very concept of science might change, so that hypothesis generation and testing, instead of remaining the paradigm cases of scientific practice, could well recede in epistemic significance to that of measuring and meter reading. What, then, would be the new exemplar of scientific activity?

If we want something distinctly human to fill this need, we should look for a mode of thought whose regularity would undermine its purpose and hence would be undesirable (even if possible) to treat in computational terms. Psychologically speaking, this means a cognitive process whose performance quality would not be affected by having a faulty memory, even though the identity of the particular outputs generated by that process almost certainly would be different. Sociologically speaking, it implies a pattern of opinion formation in which the degree of convergence on an opinion does not contribute one way or another to its credibility. These hints point in the direction of what has traditionally been called the "imagination," the seemingly random combination of ideas taken from experience. In the case of science, the relevant "combination of ideas" might be the random bibliographic search of one's own memory for texts, which when juxtaposed open up a research project.

I must admit that even a suggestion as sketchy as this does not preclude a computer prosthesis, namely, one designed with cleverly arbitrary ways of browsing through a library catalogue or lacing together a few citations. Moreover, there are classical precedents for the computer practice in the "arts of memory," whereby a rhetorician would learn mix-and-match strategies for arriving at impressive things to say to a particular audience (Ong 1958). Even the first textbook on method to show the hand of Cartesian influence, Antoine Arnauld's *The Art of Thinking* of 1662, imports the rhetorical distinction between invention and instruction (or persuasion) to discuss what we would now call, respectively, the contexts of discovery and justification in science (Arnauld 1964, part 4). Unfortunately, from the standpoint of developing option (y), the rhetorical heritage vanished with the onset of mass printing and a new sense of what personal memory is for, namely, not to *complement*, but to *internalize*, the things that lie outside itself. Crudely put, whereas before Gutenberg, students committed Aristotle to memory as an *epistemic* imperative, since they lacked personal copies of the text, afterward, with copies on hand, Aristotle was committed to memory as a *moral* imperative, to show the quality of the student's mind. In this respect, Marshall McLuhan (1965) bears close re-reading.

Therefore, what is the picture of the *human* knower that emerges from our forays into computationalism? Someone who has been designed with a set of flawed and entangled programs, which over the course of history are unraveled and alienated to the various computing devices that can do a better job. Carnap's likening of inductive logic to a Swiss army knife is instructive here, as it also serves as a model of the human being in general. Although the Swiss army knife performs each of a variety of tasks only crudely, it is perhaps the best single instrument for performing *all* of them (Margalit 1986). No doubt, for any discipline, this all-purpose inductive logic could be outperformed by an expert system loaded with the sort of domain-specific data and heuristics on which practitioners of that discipline routinely base their epistemic judgments. But only an unwieldy collection of expert systems could outperform the inductive logic in all domains. And once it is granted that the knowledge enterprise is a fallible and corrigible pursuit that ultimately aims to unify diverse bodies of empirical knowledge, then expediting the search by adopting the simpler methodology—the inductive logic—is worth the somewhat increased likelihood of erroneous judgments.

CHAPTER FOUR

Reposing the Normative Question: What Ought Knowledge Be?

1. Knowledge Policy Requires That You Find Out Where the Reason Is in Knowledge Production

The practical aim of social epistemology is "knowledge policy making" (Fuller 1988b, Appendix C). This means that the social epistemologist has an interest in "rationalizing" the knowledge production process, a task that involves locating the sorts of things that can be rationalized. Once the problem of identifying the units of rationality is set up in this way, then the tenor of philosophical argumentation starts to change. For example, faced with the several celebrated cases of fraud in the annals of science, a philosopher may be initially inclined to recommend that experiments be more closely replicated. Clearly, such a policy would *in principle* cut down the level of fraud and would, moreover, strengthen what many philosophers have regarded as the backbone of the scientific method, inductive testing. Yet, there is little reason to think that the policy, stated this baldly, has any chance of success. Indeed, the norms of formal scientific communication are themselves a major contributor to the perpetuation of fraud, as they encourage a canonical (and hence idealized) presentation of experimental procedure, combined with an imperative to publish only novel results (cf. Collins 1985). The philosopher turns social epistemologist, at this point, once she works these facts into her prescription, realizing that to recommend the replication of experiments is really to recommend a substantial reorganizaton of the normative structure of science.

Of course, this is not to deny the ultimate desirability of replication, but now the philosopher must make the case for its desirability on something more than purely conceptual grounds. Specifically, by increasing the probative burden on the defender of replication in this fashion, an issue is highlighted that has typically escaped philosophical notice: namely, whether the end that would be achieved by replication could be achieved more "economically" by some other means—where the measure of "economy" is the extent to which the greatest improvement in knowledge products can be bought at the least cost to the existing structure of knowledge production (Fuller 1988b, Appendix B). This question is motivated by two considerations, both of which point to a difference in the conception of *rationality* presupposed by the social epistemologist (i.e., one of systemic efficiency) and the more classical philosopher of science (i.e., one of methodical self-consistency).

The first consideration calls into question philosophical views about science's autonomization from the rest of society during the modern period (e.g., Shapere 1984). Contrary to the spirit of these views, it is clear that every scientific practice is simultaneously an instance of other institutional practices. For example, the act of proposing an hypothesis for testing is, at the same time, a request for capital and labor, an exercise of authority over a contested field, the maintenance of disciplinary boundaries, the reproduction of general linguistic rules—as well as a prediction of how nature will behave under specified conditions. Indeed, if the scientist cannot attend to any of these levels on demand (e.g., from a skeptical colleague who asks, "How much will it cost?"), then she is likely to be interpreted as not having proposed the hypothesis "seriously." Yet, even if taken seriously, the scientist may generate untold consequences from this initial action as its various levels interact in complex ways. Let us say that the scientist obtains a generous research grant. Her successful mobilization of resources enables extensive testing, which, if it results in early disconfirmations, may cause research in the field to close down prematurely. Needless to say, the scientist must make trade-offs from among the institutions represented by these levels, which are usefully seen as being "packed" into a finite chunk of space–time (Giddens 1984, ch. 3). Robert Merton (1976) has termed the psychology of managing this sociological scarcity "ambivalence," which the scientist feels, say, when she realizes that the methodologically soundest way of testing her hypothesis would also put her at the most risk of incurring the wrath of her colleagues.

The second consideration calls into question whether we should take philosophers of science at their word. Although they typically say that their normative theories of methodology and rationality are meant to regulate the activities of individual scientists, as was noted in Chapter Three, the theories that they actually propose seem better suited to a scientific community. For example, it would not be too far-fetched to read Popper's (1963) strategy of conjectures-and-refutations and Lakatos's (1970) methodology of scientific research programs as rather abstract blueprints for dividing up the scientific labor, namely, specifications of the sorts of labor that need to be done. As a matter of fact, their accounts are no more abstract than Talcott Parsons's (1951) systems–theoretic analysis of the social order, which identifies structural variables—what would normally be called "practices"—that are designed to perform certain essential social functions. Although the association with Parsons is hardly designed to bolster the credibility of Popper and Lakatos, it nevertheless suggests that when philosophers talk about a methodologically sound practice like replication, they may be referring not to something that an individual scientist would routinely do (i.e., a Parsonian "structure"), but rather to a systemically beneficial consequence (i.e., a Parsonian "function") that scientists unintentionally produce in the course of doing other things. In that case, the structural variables that are capable of performing the function of replication are bound to be subtle: For example, although scientists are encouraged to publish only novel results, still the need to establish priority over competitors leads scientists to shroud their research in a certain amount of secrecy, which in turn suggests that much of the same empirical ground will be unknowingly traversed by competing research teams. This unwitting duplication serves the function that would be more directly served by replicating results. An even more interesting possibility, following Brannigan and Wanner (1983), and illustrating the sorts of trade-offs discussed in the first consideration, would be that since lack of communication among scientists often accounts for multiple discoveries, if these discoveries are counted as replications, then improving scientific communication may unwittingly result in a decline of replications.

Since philosophers have traditionally been allowed to speculate from within a socially frictionless medium (i.e., "the language of thought"), they have tended to run together the ends of knowledge with the most explicit and direct means of achieving those ends. Indeed, in most cases, it is just this blurring that distinguishes the philosopher's normative conceptions of reasoning from the more circuitous routes that scientific reasoning actually takes. For example, if a scientist agrees with Popper that the end of knowledge is to eliminate error, then Popper would have the scientist adopt the most direct means toward that end, methodological falsificationism—as if commitment to an end also committed one to pursue it as an end-in-itself, which is to say, in isolation from other ends. Not surprisingly, then, philosophers have conceived of scientific methodology as categorical imperatives, which, once identified, are to be followed regardless of field of inquiry, material cost, collegial opinion, or even short-term track record at getting results. Max Weber (1954) canonized this transfer of Kantianism from ethics to epistemology in "Science as a Vocation." In turn, those who refuse to give blind allegiance to method have been portrayed as epistemic libertines, Paul Feyerabend (1975) being the most self-styled case in point. I will take up the political side of implementing methodological norms later in this chapter.

In light of the two considerations raised above, the social epistemologist cannot accept this image of science as a Manichaean universe torn between saints and sinners. But at the same time, she simply cannot ignore the concept of methodology that has given rise to the image, if only because that concept has become integral both to ordinary accounts of knowledge production and, so it would seem, to people's actual experience of producing knowledge. The strategy, then, is to establish that appeals to methodology do little to explain what scientists are doing. The point is made in two steps: (1) Show that the behavior of scientists in their "natural setting" (i.e., the laboratory workplace) is inconsistent with their self-avowed methods. (2) Show that the manufactured settings (e.g., a psychology experiment) in which scientists can be made to act methodically bear little resemblance to the scientists' natural setting. Already implicit here is one—though by no means the only—division of labor between the ethnographic approach associated with social constructivism (step 1) and the experimental approach of the recently developed cognitive psychology of science (step 2) (Tweney et al. 1981, De Mey 1982, Gholson et al. 1989, Fuller et al. 1989). A neglected but noteworthy aspect of this division is that the sociologists who conduct step (1) studies typically draw quite different conclusions about the nature of scientific reasoning from the psychologists who predominate in step (2) studies. The point deserves close scrutiny by the social epistemologist, who from her own analysis of the situation, must ultimately arrive at policy recommendations for the conduct of science.

2. Unfortunately, on This Issue, Philosophers and Sociologists Are Most Wrong Where They Most Agree

It is no secret that the attention given by philosophers of science to social constructivism is largely the result of the explicit challenge that the sociologists have made to the very existence of scientific methodology as traditionally described by philosophers. One consequence of the sociologists being dialectically poised in this manner is that they tend to portray scientific reasoning in terms that are exact complements of the ones used by philosophers. Whereas philosophers have traditionally sought the foundations for such reasoning in its global and objective character, the sociologists have argued that the rationality exhibited by scientists in their activities is best understood in terms of its local and self-interested nature. The ensuing debate, which pits the philosophers' methodologically steadfast scientist against the sociologists' opportunistic scientist ever attuned to her environment, has taken on a subtle moral tone (Fuller 1988b, ch. 10). However, from the standpoint of many psychologists who have studied scientists in simulated reasoning tasks, the striking feature of this debate is the extent to which the philosophers and sociologists are willing to presume that scientists are indeed "rational" in some appropriately defined sense, about which they then proceed to argue. The notion of rationality that provides common ground for the philosopher and sociologist is simply the idea that the scientists succeed at what they are trying to do most of the time or at least fail in ways that permit them to continue and improve upon their efforts. The bone of contention is over how scientists manage to do this, and whether we ought to approve. Yet, this seemingly harmless and intuitively acceptable presumption of scientific rationality is exactly what the psychologists contest.

As observed in Chapter Three, experiments purporting to refute one or another account of scientific rationality have come under heavy attack from all quarters: An example, allegations that the experimental subjects are only sometimes scientists, more often they are "analogue populations" drawn from undergraduate science students; that the experimental tasks simulate only the formal structure of scientific reasoning without introducing the domain-specific content that is crucial to the adeptness of real scientists; that an optimal way of addressing the experimental task is typically presupposed, which goes against the open-ended nature of actual scientific inquiry. And the list goes on. Nevertheless, at the very least, the disdain that the psychologists display toward the rationality debates that so absorb philosophers and sociologists of science alerts us to some potentially problematic assumptions. The three assumptions enumerated below are made especially by sociologists but increasingly by philosophers who have come to be influenced by social constructivist considerations. The contrary opinion following each assumption is informed by psychologistic doubts about human rationality.

(a) Scientists are more likely to be good at carrying out strategies designed to maximize their own interests than at carrying out strategies designed to produce interest-free knowledge.

(a′) On the contrary, the fact that scientists are motivated by self-interest does not ensure their competence in the conduct of science. After all, as the expected utility formula makes painfully clear, there is a strong cognitive dimension to the pursuit of self-interest, namely, the calculation of probabilities for the relevant possible outcomes that would issue in states of pleasure or pain. Moreover, if anything, long-term self-interested pursuits may well turn out to be especially oblivious to failure, as interests are continually adapted to match what can be reasonably expected (cf. Elster 1984b).

(b) Scientists can improve their performance by "learning from experience" in the research environment.

(b′) On the contrary, a robust sense of learning from experience happens only in highly controlled settings, where the scientist's behavior is subject to immediate and specific feedback. A less controlled environment, such as an ordinary research setting, is subject to irregular feedback, which makes the detection and correction of error difficult, if not impossible (Brehmer 1986).

(c) If the interpreter has problems in rendering the scientist's behavior rational, but no one in the scientist's company has such problems, then the interpreter has failed to factor in the role that "context" or "background knowledge" plays in understanding as it occurs in natural settings.

(c′) On the contrary, "context" and "background knowledge" are used so elastically as to elevate the communicative powers of the "tacit dimension" to a form of social telepathy. Indicative of this problem is the absence of agreed upon rules for when context can and cannot be used to license inferences about the content of scientific communication; indeed, context is typically whatever the interpreter happens to need to presume in order to make sense of the particular scientists under scrutiny. Appeals to "background knowledge" work in much the same way, differing from "context" only in that the former expression suggests that what is missing is to be found in the scientists' heads, whereas the latter implies that it is a feature of the scientists' common environment. By trying ever so hard to render the scientists rational, interpreters obscure a psychologically more realistic possibility, namely, that what is not said may not have been thought or even noticed, thereby enabling long-term misunderstandings to persist among scientists (cf. Fuller 1988b, ch. 6).

Whereas (a') and (b') may be generously interpreted as empirical checks on certain habits of thinking common to social constructivists and their social epistemological well-wishers, it is difficult to see (c') as anything other than a frontal assault on the very idea that knowledge is social. For what would normally pass as the unproblematic medium of social life—context—is made out to look more like the aether, that is, a protean creature of hermeneutical adhockery and rationalization. At the very least, (c') highlights the extent to which social constructivists are willing to be methodologically flexible in describing scientific practice, just in order to preserve an essentially *normative* picture of scientists as locally rational agents. However, the normative dimension of constructivist interpretations is often obscured by appeals to participant-observation as a self-certifying method. That is, if the interpreter bothers to spend some time in the laboratory, she can supposedly *see* the scientists make sense of each other and their environment in the opportunistic ways that social constructivists describe (cf. Knorr-Cetina 1980).

At best, the bromide of "seeing is believing" scores a rhetorical coup against philosophers who have never seen scientists in action and hold, with Popper (1963), that science is simply philosophy conducted by other means. But the question that begs to be asked here is when does the interpreter decide that she has seen enough of the lab habitat and that it is time to start writing what she has seen (cf. Clifford & Marcus 1986). After all, the nineteenth century anthropologists who were most likely to regard the natives as subrational were the ones who observed their practices up close and refused to abstract very much from what they saw. By contrast, anthropologists who held that reason was a universal human faculty tended to base their case on more "armchair," indeed a priori considerations, which involved abstracting away from first-hand reports (Stocking 1968). No doubt, today's interpreter would chastise her nineteenth century precursors for not having stayed long enough to see the reason in their acts.

3. However, Admitting the Full Extent of This Error Suggests a Radical Reworking of the History of Science

But now, might not this same interpreter accuse our experimental psychologists of having lingered for *too long* in the native culture, observing their actions *too closely*, and as a result, finding discrepancies between word and deed that the natives themselves fail to notice? And if the natives are not bothered, then why should the interpreter be? For several reasons. First, just as local practices that are inexplicable in their own terms may serve "latent functions" in maintaining the overall social order (Merton 1968), so too practices that make sense to the locals may be latently dysfunctional at a more global level (Harris 1974). Second, since the natives have never studied themselves with the analytic acuity of the interpreter, they may just be sufficiently impressed by the interpreter's findings that they will restructure their behavior so as to minimize its dysfunctional consequences. Third, the discrepancy highlights the looseness of fit between so-called pragmatic and epistemic virtues, by challenging the widespread intuition that the natives would not be able to survive if most of their beliefs were not true (cf. MacDonald & Pettit 1981). The most obvious source of large scale error is the failure to attend systematically to the degree of correspondence between beliefs (expressed as sentences) and reality (expressed as states of affairs). Given the urgency of the practical tasks in which the natives are normally engaged, they must simply presume that most of their beliefs are true. The interpreter, by contrast, can give the epistemic issue her undivided attention.

However, a more subtle, and philosophically more significant, source of error is the possibility that the native language makes categorical distinctions that reality does not recognize as making any essential difference. The error is most readily seen in the sorts of beliefs that an interpreter is likely to characterize as "superstitious": For example, the native believes that her tribe's crop yield depends on whether the entrails of the sacrificial bird is one color rather than another, when, as the interpreter well knows, the causal order bypasses this color test in determining the fate of the crops. I will show how an analogue to this far-fetched example may be observed in scientific practice, starting with an argument that instrumentalist philosophers of science (e.g., Laudan 1984, ch. 5) have traditionally used against scientific realists.

Suppose that a scientist has managed to set up a crucial experiment to test two theories by finding a situation where one theory predicts an effect and the other predicts the absence of that effect. The experiment is performed, and the effect is observed. The scientist then concludes that the first theory is correct, presumably because it came closer to identifying all the causal mechanisms at work in the situation. At least, this is what the scientific realist would have us believe. In response, the instrumentalist wants to deflate the epistemic significance of this "inference to the best explanation" as being nothing more than a case of affirming the consequent, since the scientist has yet to eliminate all the possible alternative theories, whose quite different causal mechanisms could have also predicted the effect. The fact that the scientist cannot conceive of these alternative theories cannot weigh in favor of the scientific realist, who holds that the nature of reality is independent of our cognitive capabilities. Thus, aside from committing an elementary logical fallacy, inference to the best explanation makes the plausibility of scientific realism turn on our lack of imagination (cf. Duhem 1954).

Her success at deconstructing the scientific realist argument notwithstanding, the instrumentalist still shares an assumption with her opponent that needs to be overturned before the potentially superstitious character of scientific practice can be seen. For the instrumentalist still assumes that one of those possible theories is right; she only doubts our ability to determine which one it is. In fact, the instrumentalist's doubts are typically so trenchant that she wants to ground the rationality of the scientific enterprise on something other than search for underlying causal mechanisms, such as the reliable generation of empirical regularities. However, the instrumentalist would be no less comfortable than the realist with the idea that there is no way of deciding which possible theory explains the experimental effect because reality itself is, so to speak, more *coarsely* grained than our ability to make discriminations, even as measured by the number of alternative theoretical languages that are *actually* available. The idea, then, would be that through a kind of *linguistic hypertrophy* (or "language going on holiday," as Wittgenstein would say), we have outsmarted ourselves into thinking that reality runs *deeper* than our language, when in fact it is the other way around!

Consider the consequences of such a view and how they might inform the observation of scientists in their habitats. Instrumentalism has been effectively turned upside down. The problem of knowledge is not, *pace* Quine (1960), that reality is *underdetermined* by our theories; rather, reality is *overdetermined* by them. As a result, in order for our theoretical distinctions to make any real difference, we need to manufacture some additional reality, ways of institutionalizing alternative causal trajectories for the alternative theories that would otherwise be so much idle puffs of air. For example, we may associate each theory with a well-defined social group that stands to gain or lose disciplinary credibility; we may conventionally tie the fates of other, causally irrelevant theories to the acceptability of a given theory; and so forth. It may even be that these social constructions enable the creation of new experimental effects, which are then taken to be signs of new entities (Hacking 1983). In any case, they make it meaningful to defend one or another theory, while at the same time subtly shifting the locus of the debate away from the events in the lab, where there is really nothing to be settled. Aside from the obvious sociological precedents for this view (esp. Harry Collins 1981), it may be found in a strongly ontological version of positivism ("phenomenalism," cf. Harre 1972, ch. 3), whereby theories of equal empirical adequacy are merely notational variants of the same theory, that theory being about nothing more—and nothing less—than the phenomena deducible from the theory. Yet, as the positivists are quick to add, these phenomena are not the only ones we should notice in distinguishing among the theory's notational variants, for the variants also evoke different behavioral responses in different people (Ayer 1971).

Admittedly, twentieth century positivists have generally cast their critical gaze at theories that aim for a higher-order, synthetic understanding of the social or natural worlds, such as political ideologies and theological doctrines. For example, the positivist observes that theologians differ over the nature of the Divine Plan, but they do not question whether the very idea of such a plan might superstitiously impute more unity to nature than it really has. Likewise, the careful observer of scientists at work may be able to spot linguistic hypertrophy going in the opposite direction, namely, toward a lower-order, ever more analytic understanding of the social and natural worlds, a search for the ultimate units of matter, life, and mind (cf. Redner 1987, esp. ch. 3). In fact, this would be more in keeping with the spirit of nineteenth century positivism, though the issue becomes considerably more vexed. A natural analogue to the linguistic hypertrophy thesis we have been examining is the Newtonian pairing of the infinite divisibility of geometric space (cf. our seemingly unlimited ability to make verbal distinctions) with matter having an indivisibly smallest unit, or "atom" (cf. the grain inherent in the structure of reality). Positivists opposed this view on two rather contradictory grounds: On the one hand, atom-talk was a theoretically useful fiction that had been illicitly turned into a imperceptible level of reality; on the other, atom-talk proposed to set an arbitrary limit on the lowest level at which it was suitable to analyze matter (why not *sub*atomic talk?). Clearly, only the former criticism makes linguistic hypertrophy out to be a research liability rather than a heuristic (cf. Capek 1961, chs. 4–7).

Needless to say, a very interesting subversive history of science could be written on the theme of linguistic hypertrophy. A good place to begin getting a grip on this issue is with Niels Bohr's explanation of the "naturalness" of Newtonian mechanics in terms of its conceptual fit with the "midworld" defined by perceptual realism: Once we try to understand reality at a finer grained level of analysis, Bohr argued, indeterminacy enters, and not surprisingly, empirical inquiry into the "subworld" has historically coincided with the emergence of physics as an academic discipline, whose methodological conventions have given structure to what would otherwise be a structureless pursuit (cf. Feyerabend 1981a, ch. 2, for a critique). The question that would ultimately loom in this subversive history is (how would one know) the point at which scientific language matched the natural grain of reality—before overshooting it!

4. But It Also Means That the Epistemic Legitimacy of the Interpretive Method Has Been Undermined

Charges of first-person superstition do not apply only to anthropologists and sociologists who purport to interpret contexts. Such charges extend equally to philosophers who follow Wittgenstein and Quine in officially suspecting that there is nothing more substantial to mental states (i.e., something intrinsic to the person to whom such states are attributed) than their contexts of attribution, but who at the same time argue that we make sense of another person's behavior by imagining what we would do under relevantly similar circumstances and then using those judgments as the basis for rendering the agent under scrutiny rational (e.g., Stich 1982, Dennett 1982). Given the authority that this sort of schizoid epistemology has commanded over the past century, first as *Verstehen* and now, more generally, as the ethnographic or hermeneutical method, it is important that we reveal the duplicity of the appeal to "interpretation."

The epistemic legitimacy of the interpretive method rests on several social psychological assumptions (Wyer & Srull 1988). These may be cast as certain epistemic relations that "we" as observers stand to an agent under scrutiny, to wit: that there are determinate answers as to what we would do in various situations, that we have especially reliable knowledge of the answers, that our actions in these situations would conform to our own theory of rationality, and that our actions are reliable indicators of what the agent under scrutiny will do. However, these assumptions run afoul not only of the constructivism that its philosophical adherents officially espouse but also of the evidence provided by psychologists. For example, if there are determinate answers to what we would do in various situations and we still want to remain consistently constructivist, then these answers must pertain to likely community attributions, given the most likely displays of behavior in those situations. In that case, there is no good reason for thinking that personal experience would be any more reliable as a source of this information than, say, the behavioral regularities derived from a social scientific study of the community. This is not to deny, of course, that we have easily accessed opinions on what we would do; however, there is no good reason for thinking that these opinions are representative of our actual experience, as opposed to being merely the reports that we have learned to give in response to questions about what we would do in various situations.

But once we grant that a social scientist's third-person empirical analysis may be a reliable source of information about our likely actions and the mental states likely to be attributed to them, then it is not so obvious that such an analysis would conform to our own theory of rationality. After all, it may be that we would perform in a way that the community would regard as suboptimally rational, but that would, with luck, be tolerated, excused, or simply not noticed. And, finally, where we tend to fall statistically within the range of rational performers in the community probably stands in no systematic relation to where the agent under scrutiny would most likely fall.

At many stages in the interpretive process, then, the interpreter's uncritical reliance on the first-person perspective leads her to reify a vividly experienced, but nonetheless faulty, understanding of the social facts pertinent to her own situation, which is unfortunately then used as the basis for "contextualizing" or "empathizing with" the agent under scrutiny (cf. Ross 1977 on the role of "availability heuristics" in everyday life). It would appear, then, that the rationality exhibited by scientists in the workplace is little more than a projection of the sociologist's uncritical appraisal of her own irrationality. The strategy needed to counter this lack of self-criticism is the promotion of an ethnographic method that does not direct the observer to render the behavior of the natives more coherent with each additional piece of data she receives. For, just as the natives should not be presumed to be any less rational than the observer, neither should they be presumed to be any more so, which seems to be the bias at work these days (cf. Fuller 1991a).

Moreover, *pace* Geertz (1973), the natives should not be presumed to find each other's behavior meaningful at a finer-grained level of analysis than members of the observer's culture find their own behavior. Indeed, the efficacy of the observer's third-person perspective lies precisely in the fact that people do not normally monitor each other's behavior very closely, as witnessed in the tolerance for alternative ways of performing in a given situation, *especially* the tolerance that the natives show to the participant–observer anthropologist! In turn, this becomes the unwitting basis for change in the culture's practices. Very often, I fear, the purported "uniqueness" of a culture's activities is an artifact of the observer making the fallacious assumption that any behavioral difference that she perceives among the natives is likely to be perceived as a meaningful difference by the natives themselves (cf. Berkowitz & Donnerstein 1982, which makes this point for interpreting social psychology experiments). There is considerable irony in this conclusion, since it implies that the very attempt on the part of ethnographers to capture a culture "in its own terms" may reflect a trenchant ethnocentric bias, which illicitly projects onto the natural attitude of the members of the alien culture the Western observer's experience of the exotic. This projection is an artifact of *relativism*, that distinctly Western way of perceiving the non-Western, whereby alien cultures are portrayed as having a stronger sense of normative bounds (e.g., sacred vs. profane space, cf. Douglas 1966) than Western culture, which, in turn, enables the members of those cultures to attach more significance to each other's behavior than the Western observer does to her own. I submit that an analogous form of relativism is at work in the humanist-based ethnography used for studying scientists, who are also said to embody a particularly heightened sense of what is inside and outside their culture, an image that is reinforced by localizing the field of study to a laboratory (cf. Bloor's [1983, ch. 7] adaptation of Douglas' grid-group analysis to Science & Technology Studies). Of the original team of lab ethnographers, Bruno Latour and Steve Woolgar, Latour has freely admitted this point (cf. Latour & Woolgar 1979, p. 18), whereas Woolgar (1988b, chs. 6–7) has come to disown it.

5. Moreover, the Fall of the Interpretive Method Threatens the New Cognitive History of Science

Our discussion of "the interpretive method" has so far been presented as if only anthropologists, with their first-hand knowledge of the natives, could practice it. But clearly, humanists have been interpreting texts for much longer. Indeed, historians of science have recently begun to naturalize the use of the interpretive method in their research by importing the cognitive sciences, thereby continuing Dilthey's original project of forging the historian's links to the past on the basis of transhistorical psychological principles (cf. Fuller 1988b, ch. 5). Yet, although much can be said in favor of treating historical texts as social-psychological data, that would nevertheless require challenging many of the old *geisteswissenschaftlich* methodological assumptions that continue to haunt the new cognitive historiography of science. I will focus my fire primarily on the work of the psychologist Ryan Tweney, whose group at Bowling Green State University set the first comprehensive research agenda for the psychology of science (Tweney et al. 1981; a still more comprehensive one is Gholson et al. 1989). As will become clear, my main quarrel with Tweney is that he should have imported *more* of psychology, not only its theories, but its *methods* as well.

In fixing the proper role of psychology in the historiography of science, Tweney (1989) realizes that "the cognitive paradigm" (De Mey 1982) enters the field fairly late in the game, since a Freud-inspired "psychohistory" of science has now enjoyed notoriety for several decades (e.g., Feuer 1963). However, Tweney argues that cognitivism has the advantage of being "closer to the data" than the psychoanalytic approaches. That Tweney should choose to pitch the difference between cognitivists and Freudians at this level is already to court trouble. If we take these "data" to be texts that arguably document a scientist's mental processes (e.g., diaries and notebooks), then it is important to realize that cognitivists and Freudians have tended to turn their analytic gaze to data from quite different periods in the history of science. Psychoanalytic methods have been most thoroughly used to study scientific figures prior to the Enlightenment, especially during the Scientific Revolution (Koestler 1959, Manuel 1969). By contrast, cognitive approaches tend to be applied to post-Enlightenment figures (e.g., Holmes 1984, on Lavoisier), and especially nineteenth century ones (Gruber 1981, on Darwin; Nersessian 1984, on field physicists), and early twentieth century ones (Holton 1978). Admittedly, there are a few exceptions on both sides (e.g., Feuer 1974, a Freudian study of the Einstein–Bohr generation; De Mey 1982, ch. 10, a cognitivist study of Harvey), but they are more speculative and less data driven. A sign of the problems to come is that a cognitivist historiography seems especially apt for scientists who make something like the internalist–externalist distinction in their own minds and are thus prone to include only what we now normally regard as "relevant" or "disciplined" thoughts, even in their private writings.

The big methodological trouble lurking here is that the scientists studied by each set of historians may have *wanted* to be interpreted roughly in the image of the historian's preferred psychological theory (though, clearly, the scientists and the historian would not be attracted to this image for the same reasons). The problem is one quite familiar to literary critics, namely, how the scientist "inscribes herself" in her text. A psychologist might call it the subject's "metacognitive" attitude, that is, her attitude toward the verbal report she gives the experimenter. More mundanely, aside from conveying explicit information about the external world, which may or may not reflect how the world really is, authors also convey implicit information about themselves, which may or may not reflect how they really are. These modes of self-presentation appear as writing and speaking conventions, perhaps even entire genres, that often tell more to the knowing reader than what is presented as the text's content. In particular, a reader familiar with the genre in question can estimate the likelihood that a line of reasoning in the text mirrors a genuine mental process, as well as the likely sources of the distortions or "filters.

The upshot so far is that, taken at face value, Tweney's claim that the cognitive approach is "closer to the data" than other historiographies of science may count as a strike against the approach, given the analytical naivete that this claim suggests. However, in all fairness to the cognitivists, they are not alone. Psychoanalytic histories typically thrive on notebooks, such as Kepler's or Newton's, that evoke mythical causal agencies alongside of real ones, and run together seemingly incongruous religious, scientific, and personal themes, since these textual features make it seem as though the scientists were recording their free associations, the mark of an unfettered unconscious, wherein supposedly lies the source of the scientists' creativity.

Needless to say, these people of the seventeenth century did not see themselves as writing in order to release their pent-up libidinal impulses. Still, a measure of epistemic authority could be garnered, at least in certain esoteric circles, by presenting oneself as a receptacle for the passage of various cosmically informed spirits. A convincing presentation of this sort required writing as if another person, someone of synoptic vision and cryptic expression, were speaking through the author. This is not to say that the author would always put on a convincing show. After all, many of these scientists were persecuted as heretics. But even the charge of heresy presupposes that the audience understood exactly (and hence disapproved of) what the author was trying to do. The Freudian reading also plays neatly into this mode of self-presentation by presuming that there is a certain way to write when one is not in full control of what is being said. The problem, I fear, is that whereas the seventeenth century reader would have easily recognized the connection as a conventional one, the Freudian psychohistorian may well assume it to be a natural sign of unreason—that Kepler and Newton actually wrote in the sort of rationally diminished state that permits a glimpse at their creative unconscious (cf. Hirst & Woolley 1981).

To get a sense of where the cognitivist is prone to commit comparable errors, consider Tweney's own reading of Michael

Faraday's voluminous laboratory notebooks. Faraday is the master of another mode of self-presentation that easily gives the impression that he is transparently recording his mental processes. As we just saw, one mode is to write so as to suggest free recall; another, quite opposite way is Faraday's, namely, to write so as to suggest scrupulous attention to detail. In fact, Faraday's notebooks are among the few privately kept scientific writings that conform to philosophical expectations about the rational conduct of inquiry. As Gorman and Carlson (1989) have observed, Faraday is the perfect Lakatosian. His research program is designed to arrive at a unified theory of the forces of nature, with a unified account of electricity and magnetism functioning as the cornerstone. True to Lakatos, Faraday is metaphysically inspired (by Schelling's *Naturphilosophie*), which leads him, in the early stages of his inquiries, to stress hypothesis confirmations over disconfirmations. However, once he had established a firm experimental track record, Faraday then proceeded to eliminate alternative hypotheses, eventually arriving at an explanation of electromagnetic induction in terms of "fields of force." In addition, the notebooks reveal that Faraday regularly recognized and corrected past errors in theory and method. All these features straightforwardly suggest to Tweney that Faraday successfully employed a set of problem-solving heuristics that has been isolated by psychologists and artificial intelligence researchers. However, there are several historiographical problems with admitting this conclusion so quickly, problems that we will now examine in turn.

First, the cognitivist approach commits its own version of the old hermeneutical fallacy of assuming that what transpires between the covers of a volume (or set of volumes, in Faraday's case) is to be interpreted as a *coherent* work. The cognitivist version is to treat a scientist's corpus as an extended exercise in problem-solving, which suggests a pre-existent goal and the directing of thought toward that goal, usually in the form of constraints on permissible solutions. True, the nature of the problem may change in midstream, but only if the problem-solver explicitly makes such a change. This model would make the most sense of Faraday's situation had the historian originally programmed the nineteenth century scientist to unify the forces of nature but had since forgotten which heuristics were built into The Faraday Program. However, it goes without saying that the historian did not design Faraday, yet the illusion may be maintained in several ways:

(d) Assume that, in the absence of external interference, once Faraday no longer revises a conclusion, he has reached an adequate solution to a problem he has been working on (for a critique of this sort of inference, cf. Rich 1983).

(e) Assume that the feeling of purposefulness that accompanies all of Faraday's thought (i.e., the "determining tendency" first isolated by the Wurzburg psychologists; cf. Boring 1957, ch. 18) implies that all of his thought has a purpose that guides its course.

(f) Include only cognitive biases that facilitate Faraday's problem-solving (i.e., heuristics) and exclude any that could impede it (i.e., liabilities), even though the chance that a human reasoner would embody such an asymmetrical set of biases is extremely small, for, strictly speaking, a heuristic is nothing but a liability on borrowed time (cf. Hogarth 1986). (It is ironic that Tweney should adopt this line, given that his own experimental work has primarily explored the cognitive liabilities of scientists; cf. Tweney et al. 1981, chs. 18, 19, 30).

On the basis of these historiographical biases, it is correspondingly unlikely that any of the following would ever be found in a cognitivist account of Faraday:

(d') Faraday either mistakenly revises or stops just short of what the historian would regard as an adequate solution, without any signs of external interference.

(e') Faraday falsely identifies a train of thought as continuing an earlier problematic, when in fact he has significantly, though unwittingly, altered its course;

(f') Faraday hits upon what the historian regards as an adequate solution, but by the mutual cancellation of errors, which Faraday (erroneously) treats as a brilliant problem-solving strategy.

It might be objected that the meticulous detail of Faraday's diaries overrules the possibility that he could have been just as incompetent as (d')–(f') suggest. But this is to miss the point of my alternative interpretive strategies. I hardly want to deny that Faraday had an unusual presence of mind; rather, I want to deny that the source of Faraday's mental presence was necessarily any finely honed problem-solving abilities. Instead, I would argue that what is most evident from the notebooks is that Faraday is a master of the *cognitivist literary style*, his distinctive mode of self-presentation. To make the case that Faraday is an exemplary problem-solver involves the additional step of correlating the notebook entries with laboratory practices, especially with an eye to the delay between the time Faraday engages in lab work and the time he writes about it (Gooding 1985; cf. Downes 1990, for similar advice on evaluating Robert Millikan's notebooks). Otherwise, we simply have Faraday's own self-image. As Tweney himself notes, the young Faraday attended lectures given by Isaac Watts, a popularizer of Lockean empiricism, who offered counsel on improving the mind. Watts advised that a commonplace book be kept, in which ideas are recorded with the express purpose of reflecting upon them on later occasions, thereby creating a continuous train of ever more refined and integrated thoughts. It is easy to see how undertaking this disciplined form of writing would sustain the image of the rational mind at work. Just imagine the extent to which Faraday would need to edit his real thoughts before committing them in writing as logical extensions of his previous entries! Let us not forget that the main audience for Faraday's feats of self-discipline was Faraday himself, as the sight of his sustained cogitations reinforced his conviction that he was indeed headed for the truth. After all, Faraday *did* believe Watts's rhetoric. But it does not follow that the historian should as well.

Perhaps the most ironic feature of the new cognitivist history of science is that its source disciplines, experimental psychology and history, have traditionally been sensitive to the very problems of interpretation that I have been raising. Yet, this dual lineage does not seem to be enough to prevent the cognitivists from falling into the same methodological traps as internal historians of science. There is remarkable agreement among psychologists (e.g., Ericsson & Simon 1984) and historians (e.g., Dibble 1964) about the specific problems that arise in trying to treat verbal accounts as more or less transparent representations of someone's mind or ambient surroundings. Of course, none of this is to deny the central importance of verbal accounts to understanding history. However, it is a big mistake to think that the historian is in anything like a "dialogue" with the past, insofar as this metaphor suggests that, say, Faraday wrote so as to provide today's historian maximum access to his mind. Rather, the historian is better seen in the role of eavesdropper, with all the attendant interpretive risks that that metaphor entails (cf. Fuller 1988b, ch. 6). In that case, Faraday's verbal accounts should be taken as indicators of something—but of what exactly remains to be empirically determined. For a sense of the interpretive perils that await us in this task, consider a brief list of conclusions culled from the pages of Ericsson and Simon (1984) and Dibble (1964). The list should be read as a set of constraints on the psychological plausibility of any historical interpretation (cf. Faust 1984, pp. 117 ff.):

(1) You can probably give a more accurate account of what

someone else is doing (assuming that you can observe her) than of what you yourself are doing, largely because in the former case you are less likely to embellish for purposes of rendering your behavior coherent with your self-image.

(2) The measure of accuracy that you strive for in your accounts is a function of whom you are accountable to. These implicit judges of your behavior best explain why you are likely to fret over certain details and completely ignore others.

(3) Explanatory accounts of your own behavior are no less speculative than your explanations of other people's behavior, largely because the account you give is more a function of how the request for an explanation was posed than of any "natural" characterization of your behavior. In fact, by the time you are asked to explain your behavior, you have probably forgotten the way you originally characterized it to yourself.

(4) Given the initially rapid decay of the memory trace, any amount of uncorroborated detail about an event after its occurrence should be taken as products of the imagination, which may or may not bear some resemblance to the original event. This point is of special relevance, if you (like Faraday perhaps) provide relatively *long* accounts soon after the event.

(5) What you explicitly avow as your "beliefs" and "desires" are poor explainers and predictors of your behavior. Both function rather as standards by which you are likely to evaluate your behavior after it has taken place.

If (1)–(5) are not perils enough, recall that we have been taking for granted that you *are* trying to speak your mind sincerely, and not strategically distorting communication. In other words, these are cognitive liabilities built into the human organism over which it has little direct control (cf. Elster 1984a, pp. 18–22, on "cold mechanisms").

Moreover, we have yet to touch upon an even more trenchant assumption of the new cognitivist history, one which it also shares with the old internalism. It is the idea that determining the epistemic significance of a text involves little more than reconstructing the author's cognitive processes. This assumption helps explain the new cognitivist history's neglect of the psychology of idea reception (cf. Holub 1984), since reception would seem to be merely a case of the reader's reproducing for herself the author's original generation of the idea. In its more sophisticated forms, this *transparency of reception* thesis is played out in terms of two historiographical assumptions that can be detected in the interesting work of Nancy Nersessian (1984) on the development of the concept of "field" in physics from Faraday to Einstein.

Nersessian portrays four physicists (Maxwell and Lorentz are the connecting links) as successively enhancing a model whose main features are unproblematically transmitted from one scientist to the next. Nersessian is able to tell this story without saying anything about the reading and writing conventions of these scientists because she assumes (1) that generating an idea and packaging that idea for an audience form a sequence of psychologically discrete processes that correspond, roughly, to the private and public papers of the scientist, and (2) that the packaged idea reveals enough of the original author's (i.e., Faraday's) cognitive processes that later scientists are able to constitute a "research tradition" that stems from continuing the original author's line of thought. However, given the significance of these assumptions, I can only bring them to the reader's attention and defer full critique to another occasion. If the reader wants a headstart, she should look at Lindholm (1981), which criticizes internalist historians for making it seem as though there is a much smaller incidence of misunderstanding among scientists of the past than among ordinary mortals, or even scientists, of the present.

6. Still, None of This Need Endanger the Rationality of Science, if We Look in Other Directions

So far, the chances of locating rationality in science look dismal. But we need not lose hope, as long as we keep an open mind as to the exact unit of rationality. Although appeals to context in anthropology and history suggest social analyses of knowledge, the vehicles of rationality normally supported by such appeals are individual agents; hence, a shared social context is said to confer rationality on each member of a community. Still, the fact that individuals do not appear rational when judged on their own terms should not be taken as a decisive mark against there being any rationality whatsoever in the social world. After all, most standard models of rationality—whether they pertain to science or the economy—were originally proposed on conceptual grounds and survive largely on their conceptual merits, which is to say, they subsist at a level of abstraction that tantalizes us with the promise of desirable consequences (i.e., of providing the sort of knowledge or goods that we want), but with little sense of how this promise is to be empirically realized.

However, we need not remain in this paralyzed state indefinitely because there are at least three disciplines—theoretical sociology, evolutionary biology, and behavioral economics—that have confronted the problem of reason, in a world fraught with context, much more successfully than philosophers and sociologists of science. As will become clear, the key to the success of these disciplines is their explicit discussion of the vehicles of rationality, or more abstractly (to fit the biological case better), the sort of thing on which order can be conferred. Let us now briefly consider the dialectical strategies that these disciplines have to offer in channeling the scientific rationality debates in a more productive direction.

First consider the "micro–macro" dispute in sociology (Knorr-Cetina & Cicourel 1981, Alexander et al. 1986). On the one hand, the likes of Parsons have sought a general theory of society. The search has led them to rise to the level of abstraction needed to specify functions common to all societies. Given the nature of this enterprise, Parsons et al. have had to allow for many mutually incompatible practices to satisfy each of the designated functions. Even a function as basic as intergenerational survival places few a priori constraints on what may count as an appropriate social practice. For example, intergenerational survival does not preclude the ritualistic slaughter of reproductively fit members of the society. But if the range of possible practices includes even this one, what reason do we then have to think that each of these practices are somehow serving the same function in its respective society? It would seem that only faith in the macro-perspective itself—that there are societal universals to be found—would motivate the sociologist to see these palpable differences in practice as ultimately superficial.

It is here that the micro-perspective gets off the ground. In fact, the most provocative of recent micro-sociologists, Harold Garfinkel, trained under Parsons in the 1950s and soon afterward developed "ethnomethodology" as the standard strategy for reinforcing the sociologist's sense that apparent differences in practice are indeed real. It would be instructive to cast the leading candidates for "scientific universals," the criteria demarcating science from non-science, in terms of this dispute. Which practices, if any, are in principle excluded by the claim that science is an "empirical" or, for that matter, a "rational" activity? What about science itself: Are there any social practices that in principle could not count as "science?" My guess is that given these formulations of the key questions, the onus is squarely on the shoulders of the macro-metascientist, especially if she wants to argue not merely that there are common features to science but also that there has always been science. Shapere (1984) and Laudan (1983) have already given up the struggle. To counter the radical differences in practices that have passed as science through the ages (e.g., theology, philology, physics), she would need to show that these practices performed the same complex of functions in each society that was essential for their all being identified as a science (cf. Fuller 1988b, ch. 7, for some Parsons-like opening moves).

Whereas the concern in sociology was how (and whether) the need for societies to perform certain functions constrains the possibilities for practice, in evolutionary biology it is how (and perhaps even whether) selection processes constrain the possibilities for genetic variation. The analogies between natural selection and theory selection have not escaped the eye of the evolutionary epistemologist (e.g., Toulmin 1972; Giere 1988, ch. 1), except that the biologists and the philosophers focus the subsequent debate quite differently. Both disciplines have a problem with the context-sensitivity of their respective form of selection. Biologists find that a gene may be selected in certain genetic environments but not others, whereas philosophers find that a theory may be selected by some scientists but not others (or not at the same time). Philosophers typically take this state of affairs as brute and complain about the lack of appropriate (i.e., rational) uniformity in scientists' behavior. As we have seen, social constructivists also accept this as a brute description but turn an apparent adversity into a virtue, with their context-dependent notions of rationality. By contrast, biologists normally take the workings of natural selection as brute and then attempt to redescribe the situation without making reference to differences in genetic context, since selection processes operate only when the variation in the fitness of a population is heritable, which implies context-independence (Wimsatt 1984). In some cases, biologists eliminate context-talk by making the gene only part of the proper unit of selection, in other cases by showing that two separate selection processes are in fact operating (Sober & Lewontin 1984).

By analogy, then, on the one hand, philosophers could avoid context-talk by considering that something more than just a theory is being selected in so-called cases of theory selection. Although the proliferation of such philosophical entities as "research programs" (Lakatos), "research traditions" (Laudan), and "paradigms" (Kuhn) suggests that this possibility has crossed people's minds, there has been little discussion of whether analyzing the history of science in terms of one of these larger units would make the history look more rational. It may even be that a theory whose acceptance seems to vary contextually is really little more than a set of historically related homonyms that serve quite different roles in their respective research programs (cf. Fuller 1988b, ch. 9). On the other hand, philosophers might consider whether "theory selection" describes a univocal process or, rather, a host of diverse epistemic strategies that eventuate in the marked absence or presence of concern about a given theory. To his credit, Laudan (1981) has already put this possibility on the table by distinguishing the various "cognitive stances" that may be adopted toward a theory. However, it would be better if the distinction in stances were drawn from the standpoint of the likely behavioral consequences in a given domain (e.g., whether or not a particular theory is developed) rather than from the standpoint of the scientist's own perceived sense of difference. For example, while Popper (1959) and van Fraassen (1980) have attached major philosophical significance to the perceived difference between "fully believing" that a theory is true and "merely accepting" it for purposes of probing its limits, I would want to see whether this distinction issues in demonstrably different behaviors before I agree that two separate cognitive stances are at work.

To appreciate the full extent to which thinking about rationality in other disciplines has moved beyond its current doldrums in Science & Technology Studies, we need only consider the matter from the standpoint of economic rationality. When it comes to explaining economic behavior, the priority of economics over psychology has been traditionally analogous to the priority of the philosophy of science over the sociology of knowledge: The latter discipline becomes relevant only once the behavior under study (of economic or scientific agents) is demonstrably arational and hence unexplainable by the categories of the former discipline. Moreover, this order of explanation has been implicitly respected by both parties in both pairs. Thus, just as sociologists, by rendering seemingly rational cases of scientific activity arational, have defended themselves against the philosopher, we have seen the followers of Tversky and Kahneman, by showing the subtle but pervasive failures of economic rationality, try to push the case for psychologism in economics. However, this way of drawing the disciplinary boundaries has proven unsatisfactory to many psychologists and economists—if not to enough sociologists and philosophers.

Given the broad conceptual appeal of expected utility maximization as a model of rationality per se (cf. Elster 1979, part 1), more and more empirically minded economists and behavioral psychologists are taking Tversky and Kahneman's reports of the model's demise as greatly exaggerated. These inquirers have now become explicitly concerned with locating the level of behavioral analysis for which the principle of expected utility offers the best explanation (Lea et al. 1987, ch. 5, reviews this research). Most grant the trenchant character of individual irrationality for the sake of argument but then go on to test for utility maximization operating at more general, or "molar," levels of behavioral analysis, such as the intimate group, the temporally extended group, the spatially diffuse group, and so forth. A noteworthy exception in this vein has been the behaviorist Howard Rachlin (1980, Rachlin et al. 1986) who has made the cleanest behavioral translation yet of the mentalistic terms in which the rationality debates are normally couched, while still retaining the individual organism as the unit of analysis. Rachlin has argued persuasively that the irrationality of Tversky and Kahneman's subjects is an artifact of the experimenters' expecting "instant rationality" from them, that is,, subjects are made to decide between a set of options *now*, rather than decide one way now but another later, or even to postpone the decision entirely until a more opportune time is perceived.

If utility maximization is to be taken seriously, then the experimenter needs to know the exact utility that the subject is trying to maximize: What is the end toward which the subject is mobilizing her means? Answering this question is propaedeutic to making any clear sense of a subject's rationality. But the end subserved by a subject's behavior cannot be inferred, until the experimenter has had an opportunity to observe the subject adjust her behavior over time, in response to new information. As it stands, Tversky and Kahneman (and most other experimenters, who fail to take into account the decision-making history, or "reinforcement schedule," of the subject) seem to suppose that the relevant utility that will reveal the subject's degree of rationality must be one that the subject can maximize in the span of *one* decision—a gross underestimation of the subject's ability and desire to defer gratification and think in terms of long-term life plans!

With respect to the debates over scientific rationality once again Lakatos (1970) and Laudan (1977) have already made the general point that the rationality of a research program cannot be judged on the basis of one moment in its history. However, they have failed to push beyond this point in the directions that Rachlin's research suggests. In particular, a big part of Rachlin's message is that the very identity of a subject's life plan or a scientific research program cannot be specified independently of its behavioral history. This raises four questions:

(i) the extent to which subjects (in this case, scientists) need to know their goals before heading toward them, and even whether they need know them at all before admitting that they have been achieved or thwarted;

(ii) the extent to which the verbal reports of the subjects (or scientists) correlate well with the account that does the best job of making out their long-term behavior as maximizing utility;

(iii) the extent to which subjects (or scientists) ever change goals or merely postpone them indefinitely;

(iv) the extent to which subjects (or scientists) must be judged as having changed their goals when too much of their behavior starts looking irrational, even if they maintain that they are pursuing the same goals.

Since philosophers of science generally encounter the history of science with a preassigned set of labels to stick on people and movements (e.g., "Aristotelian," "Darwinian," "Newtonian"), shifting attention to these four questions would provide a healthy dose of empiricism to the rationality debates. But philosophers should also draw a somewhat subtler conclusion, namely, that many of the stock-in-trade issues of personal identity theory may be of central importance in deciding who or what can be rational in the history of science (cf. Williams 1973, Parfit 1984).

7. Reconstucting Rationality I: Getting History into Gear

In Chapter Three, we saw that the quest for a "naturalistic" theory of knowledge equivocates between two sorts of naturalism, one more attuned to the *Geisteswissenschaften* and another to the *Naturwissenschaften*. A helpful way of encapsulating this difference is to consider what it might mean to say (as Shapere [1987] does) that science should be studied "on its own terms":

(G) The study of science should capture the scientists' own level of reflectiveness about their activities, including their sense of methodological soundness and of the aims that they see their inquiries as promoting. In that case, the words of the reflective scientist, particularly when she offers reasons for accepting or rejecting various epistemic claims, are the primary evidence base for the student of science. The less reflective behaviors of the scientist, which are a matter of professional survival (e.g., grant writing) or rote habit (e.g., apparatus manipulation), enter only as indications of whether the scientist is living up to her own standards.

(N) The study of science should proceed according to the methods of science itself. Presumably, we would not be interested in studying science if we did not have prima facie grounds for thinking that science penetrates the nature of what it studies better than its competitors. In that case, the student of science takes a thoroughgoing third-person perspective to the scientist, which means that the scientist's words and deeds are treated as behaviors that may diverge or converge for reasons, and with a regularity, unbeknownst to the scientist herself. After all, the scientist under study is an expert in her field, not in her own expertise. From this perspective, the scientist is, in effect, only the site for the display of behaviors that are elicited by cues or reinforcers in her environment.

(G) is the *geisteswissenschaftlich* approach prevalent today, whereas (N) is the more *naturwissenschaftlich* variant that I prefer. Notice that in (G), science's "own terms" turn out to be those of the *scientists*, whereas in (N), they turn out to be those of the *scientific method*. One of the curious features of (G)-type naturalists is the modesty of their normative aspirations, which we originally saw in Laudan's grafting of a presentist sense of scientific progress on a relativistic sense of scientific rationality. At that point in the argument, however, I was highlighting Laudan's similarities with Lakatos, who would have philosophers evaluate past science by its contribution to present science and then recommend how the best features of present science can be extended indefinitely into the future. Laudan differs from Lakatos on just this last point, in that the output from Laudan's normative naturalism is simply a set of hypothetical imperatives about how scientists should proceed if they are in a given epistemic state and want to achieve a given cognitive aim. These imperatives summarize what has worked in the history of science, but they offer no prescriptions for the future—indeed, much in the spirit of Max Weber's steadfast social scientist who steers clear of political decision-making. Thus, in the end, Laudan relativizes even present-day concerns for knowledge by allowing the possibility that, for all their proven efficacy, many of these imperatives may turn out not to be relevant to future cognitive aims.

My diagnosis of the normative modesty exhibited in Laudan and other (G)-type naturalists is that it results from *the temporal asymmetry of rational judgment*. The "asymmetry" lies in rationality being *presumed* of the past, but *denied* of the future. More precisely, (G)-type naturalists seem to think that the activity of rendering past scientists rational by attending to "context" and "background knowledge" rests on more secure epistemic ground than the activity of rendering the future rational by designing the research environments that will produce the kinds of knowledge we will want. But as we saw in the last chapter, the first activity is fraught with epistemological difficulties that make it at least as dubious as the second—if not more so. Nevertheless, (G)-type naturalists offer a host of reasons for the asymmetry. Here are three arguments culled from Brown (1988a), Shapere (1987), and Popper (1957), which are arranged in terms of increasing asymmetry.

(h) If we study science because we presume it has been successful, then the scientists must know what they are doing. Consequently, the most that philosophers can do is to provide an account of the nature of that success, one that suffices for epistemological purposes, but one that is probably too abstract to be of much use to the scientists themselves in designing future research.

(i) Since scientists determine what science is and they have changed their minds quite substantially during the history of science, it would be presumptuous to use the past to determine the future course of science.

(j) Science is a very complicated matter, quite unlike any other social practice. Since its focus is given by something outside society (i.e., external reality), it cannot be strictly regulated; on the contrary, the point of methodology (especially falsificationism) is to prevent the most adverse effects of such regulation. Consequently, what allows us now to say that some bit of past science is good is that it has survived in several different research settings and, no doubt, several attempts at regulation. Since we do not have this vantage point with regard to the future, we have no basis for knowing what will turn to out to be good science.

Whereas (h) suggests that philosophers can at best achieve an abstract (one might say "verbal") account of what the scientist knows in practice (and hence "tacitly"), (i) suggests that this might be beside the point if scientists significantly rethink what they are trying to do, and (j) goes so far as to say that all such accounts are immaterial if science is done right, since that would mean following the truth wherever it may lead, regardless of research design. For our purposes here, the interesting feature of these arguments is the dual identity of the villain that they presuppose. The villain is clearly someone interested in planning the future of science. It is equally clear that this villain must adhere to the temporal symmetry of rational judgment, in other words, that rationality is no more vexed with regard to the future than to the past. And who is more likely to be so even-minded in her attitude toward time than our (N)-type naturalist in search of behavioral regularities and underlying causal mechanisms? But before explaining the connection that I see between (N)-type naturalism and a more robust normative orientation, a few more critical remarks need to be made about the assumptions that often make a (G)-type naturalism seem more attractive than an (N)-type.

Advocates of (G)-type naturalism tend to stress the special "complexity" of human phenomena, which cannot be experimentally isolated as variables in a closed system, a precondition (so it would seem) for testing any robust version of (N)-type naturalism. The debate surrounding this point will be played out in the next section, but something can now be said about it as a point of historical methodology. Yes, in searching the history of science for efficacious research strategies, we should consider the full range of evidence that such a history offers, since anything *might* turn out to be efficacious. However, an impartial examination of the evidence need not, and probably will not, show that all of it is equally relevant to whatever brought about an exemplary piece of research. And certainly, relevance should not be allotted in proportion to the amount of evidence available. For example, it is entirely conceivable that a historian could study Newton's voluminous theological tracts closely and conclude that they were not essential to his arriving at the formula for universal gravitation. Since the historian is trying to answer a question about causation, what matters is not how much actual time Newton spent on theology, *but how much difference it would have made to his physics had he not spent that time on theology.* Even without carrying out the relevant counterfactual analysis (cf. Elster 1978, 1983), we might surmise that one consequence of the humanities' not explicitly engaging in causal analyses is that its practitioners lack the scientist's robust sense of "wasting time" wading through piles of data, the hypothesized significance of which fails to pan out.

Sometimes, advocates of (G)-type naturalism try to obscure the fact that they are trading in causal claims by noting the "value-ladenness" of causal hypotheses. That is, we are interested in explaining, say, how Newton arrived at the formula for universal gravitation because Newton is an exemplar of our own science. Moreover, the sort of hypothesis that we will want to test (e.g., that Newton was a Bayesian methodologist) will probably reflect features that we esteem in our own scientific methodology. This orientation is clear in Lakatos, Laudan, and most normative philosophers of science. Indeed, it is so pervasive in hypothesis formation that Max Weber (1964) recommended falsificationism as a general methodological antidote.

However, the converse of the Weber thesis is probably more telling to the naturalist, for it says that all value claims are "fact-laden," more specifically, laden with causal presuppositions. Thus, to seek an explanation for something called "Newton's achievement" is already to presume certain things about the structure of historical causation, which, were they shown to be false, would probably alter the historian's interest in continuing her inquiry (cf. Finocchiaro 1973, ch. 15). One such presumption is that differences in the contributions that individual scientists make to the growth of knowledge can be explained, in large measure, by differences in the quality of their minds. However, if it were shown that Newton and other alleged geniuses had been endowed with fairly ordinary cognitive powers or that their distinctive insights were misinterpreted—or for that matter, introduced—by their successors, then the historian might legitimately wonder whether "Newton's achievement" has been properly conceived as an object of inquiry. Therefore, to bring normative naturalism full circle, it would have to be said that we value Newton as a topic for study because—and perhaps only insofar as—we regard him as having been causally significant in bringing about the science that we currently value.

The dialectical interplay between fact and value exhibited above should warn the (G)-type naturalist that (N)-type naturalists are not the sole purveyors of causal analyses. Indeed, this interplay is part of a larger complementarity that existed between the *Naturwissenschaften* and the *Geisteswissenschaften* in the writings of the Neo-Kantian philosopher Heinrich Rickert (1986), at the beginning of this century. Rickert was instrumental in completing the Millian revolution in German epistemology by reformulating the distinction between *Natur* and *Geist* as concerning, not two types of substances (as, say, Dilthey had tended to think), but two ways of studying any type of phenomena. It is at this point that we see something that resembles our distinction between (G)- and (N)-type naturalisms. The *Geist* side became the "idiographic" method, which applied equally to natural history and human history, aiming, as both were, for the unique features of their respective domains; the *Natur* side became the "nomothetic" method, which applied equally to physics and economics, aiming, as both were, for the lawlike regularities in theirs. Whereas the nomothetic method determines what must follow from a set of initial conditions, the idiographic method is left to determine which set of conditions best characterizes a particular episode under study.

Georg von Wright (1971) appreciates the complementarity implicit here when he imagines that the facts pertaining to some domain are arranged as a set of syllogisms and observes that the task of "explanation" (i.e., the nomothetic method), then, is to fill in the major premise, while the task of "interpretation" (i.e., the idiographic method) is to fill in the minor premise. The dialectical interplay between the two methods occurs over explicating the "middle term" of the syllogism, in other words, the "sense" in which the case specified in the minor premise is subsumable under the universal specified in the major premise. In brief, the nomothetist tries to subsume what the idiographist wants to deconstruct. Let us look at this interplay a bit more closely and then offer an example of how it is supposed to work.

Like most epistemologists interested in remaking the human sciences in a Millian mold, Rickert was impressed by the advances made in experimental psychology, especially its techniques for clarifying the relation between its own concerns (i.e., the mind's "internal reality") and those of the physical sciences (i.e., external reality). One such technique that captured many epistemologists' imaginations was the *difference threshold*, that is, the minimal amount by which some physical stimulus needs to be increased before a significantly changed psychological response is registered. The difference threshold was a key concept in the branch of psychology where the first lawlike regularities were obtained, psychophysics, which studied the thresholds of sensory perception, the very interface between internal and external realities. Similarly, a useful way of reading the relation between Rickert's nomothetic and idiographic methods is as a concern with thresholds, but now with the minimal amount by which our knowledge of the causal structure of history would need to be changed before our value estimation of a given event would likewise be changed. A sympathetic appraisal of the idiographic method's search for the "unique" is as a search for differences that make a difference (cf. Rickert 1986, ch. 4). But insofar as these differences are causal ones, counterfactual reasoning, and hence the nomothetic method, is also implicated. In effect, then, the differences in question are twofold. On the one hand, there is the *value difference* that minimal changes in the known event would have made to its significance for the historian; on the other, there is the *causal difference* that minimal changes in the surrounding events would have made to transforming the known event into one that the historian would regard as significantly different.

If nothing else, Rickert's historian cannot be accused of "modal naivete," since she is keenly aware of the *contingent* character of her subject matter, that is, how a minor change in the conditions surrounding an event, perhaps even in how those conditions are interpreted, may radically alter the significance of what followed. In both their never-ending search for suppressed alternative accounts of canonically defined events and their following through on the consequences of accepting any of those alternatives as canonical, the social constructivists stand virtually alone among today's students of science in doggedly resisting modal naivete (e.g., Woolgar 1988b). As opposed to their practice, we are beset by modally naive versions of the nomothetic and idiographic approaches to the history of science. Whereas some see historical cases as straightforwardly testing empirical generalizations about the scientific method and do not consider whether the very description of a given case depends on its having been read in terms of one of the competing methods (e.g., Laudan et al. 1986; criticized in Nickles 1986), others more commonly commit the opposite error of presenting a "thick description," which is often little more than a mere inventory of perspectives and details, as if that were enough to articulate the complex interaction of factors in a given historical episode. However, one current debate in science studies has exhibited the dialectical interplay between nomothetic and idiographic approaches that I have been urging. A brief look is in order.

The debate concerns the nature of scientific discovery, especially the status of so-called multiple or simultaneous discoveries. It is an issue that quickly brings to the surface deep-seated causal assumptions that are made by the various practitioners of science studies. Why did some knowledge products, such as the telephone and the energy conservation principle, emerge at several different places at roughly the same time, whereas other knowledge products had only one origin from which they then spread? Earlier in this century, philosophers took the multiplicity of a discovery to be a sign of the objectivity of the thing discovered, whereas sociologists argued that it implicated a common Zeitgeist. Despite this initial difference, both believed that multiples would increase as science progressed, either because science would gradually converge on reality (so said the philosophers) or because science would become a more institutionalized social practice (so said the sociologists). In recent years, the debate has taken a new turn, where both sides question the long-term increase in multiples. An experimental psychologist, Dean Simonton, claims that their distribution is constant, whereas a social constructivist, Augustine Brannigan, contends that the number of multiple discoveries tends to decrease over time. In any case, Simonton and Brannigan are exemplars of how to argue, respectively, the nomothetic and idiographic sides of an issue.

Simonton (1988) has proposed a "chance-configuration" theory of scientific creativity that argues for the autonomy of genius from the ambient society by pointing out that the distribution of multiple discoveries throughout the history of science closely approximates a

Poisson curve, which is what one would expect if the discoveries occurred by chance. If that is the case, then differences in social environment do not seem to constrain the occurrence of scientific discovery, which leaves ample explanatory room for a psychology of the creative mind. Brannigan (1981) has replied by contesting the instances of multiple discovery on which Simonton relies, noting that many had been first constructed as multiples by Whig historians abstracting common elements from quite disparate cognitive contexts—a point that Kuhn had observed about the celebrated convergence on the energy conservation principle.

In effect, Brannigan is deconstructing the middle term of Simonton's categorical syllogism, in the hope that if enough cases are disputed, then the truth of the major premise—that all scientific discoveries are the products of genius—will be called into question. However, Brannigan's point is not entirely negative, though he proposes a decidedly unheroic hypothesis to explain the multiple discoveries that withstand his scrutiny, namely, that they are merely symptoms of the relevant scientists not knowing of each other's work. In short, multiples are significant only because ignorance has been mystified. Over the years, the frequency of multiples has decreased, which would be expected if communication among scientists has improved (Brannigan & Wanner 1983). Needless to say, Simonton could continue the dialectic by contesting Brannigan's psychological assumptions about the relation between the accessibility to scientific work and its being accessed (cf. MacIver 1947), the relation between such work being accessed and its having demonstrable effect on one's own work, and so forth.

The discussion up to this point has served to show that the enterprise implied by (G)-type naturalism, roughly Rickert's idiographic method, can be fully realized only by incorporating the nomothetic method more appropriate to the (N)-type, which involves, by use of counterfactual reasoning, speculative experimental analyses of the causal structure of the history of science. Indeed, there is conceptual space for an experimental philosophy of science, the humanist's answer to psychophysics, which derives correlations between changes in one's knowledge of the causal structure of history and changes in the values that one confers on particular episodes in history. In a Peircean rage to neologize, we may call this pursuit, *axioaetiotics*, the study of the values–causes nexus. To conduct research in the fledgling science, the various practitioners of Science & Technology Studies (i.e., philosophers, historians, sociologists, psychologists) would be provided protocols of the form:

(P) Suppose you were to come to believe that R is the case. How would this affect your judgment about the relative importance of studying S? R = an event or fact that you do not currently believe to be the case (but will entertain for the sake of the experiment). S = an event or fact that you currently find worthy of study (but may find more or less so if R were true).

The responses would then be examined for their implicit value and causal assumptions. And even if the development of such a discipline did not end up resolving all of our value and factual disputes about the history of science, it would at least raise such disputes to a level of reflectiveness that they have yet to achieve.

8. Reconstructing Rationality II: Experiment Against the Infidels

I want now to press beyond the claim that (G)-type naturalism requires the (N)-type and argue that the (N)-type, with its reliance on experimental analysis, is potentially more valuable to the normative enterprise of the philosophy of science than the historical method of the (G)-type. My argument will no doubt strike many as highly tendentious, since the experimental science of science—in particular, the psychology of science—is still in a troubled infancy, whereas the history of science is an established resource for normative ideas in the philosophy of science. However, much of what I suspect is a priori resistance to experimental approaches stems from a failure to understand how knowledge is gained through experimentation. Given that it has now become fashionable for philosophers of science to regard "experiment as the motor of scientific progress" (Ackermann 1988), it would be both an irony and an embarrassment if philosophers were alone in their ignorance. But as it turns out, many psychologists seem to suffer from the same cognitive deficiency, which brought on the need for an excellent piece by two social psychologists, Leonard Berkowitz and Edward Donnerstein (1982), in defense of the experimental method. In what follows, I will liberally borrow from their arguments, tailoring them to the specific points that I want to make.

Let me start with an objection that the historicist philosopher of science, Harold Brown (1989; criticizing Gholson & Houts 1989), has raised against the very idea of a "cognitive psychology of science," as the objection is indicative of the conceptual obstacle that needs to be overcome. Brown observes that psychologists tend to test subjects on whether their problem-solving behavior conforms to some standard philosophical, usually positivist, model of scientific reasoning. For example, psychologists want to know when, if ever, do subjects try to falsify hypotheses, and when they are simply biased toward confirming a pet hypothesis. For Brown, however, this is already to set off on the wrong foot, since these models have been discredited by historians of science, who have shown that nothing nearly so formulaic transpires when scientists reason. On the contrary, the most decisive bits of reasoning are very much tied to the scientist's particular problem-solving context (cf. Brown 1978, 1988b).

Brown's objection seems quite potent, until we notice that it runs together two sorts of validity challenges to experimental studies on scientific reasoning:

(k) *Ecological Validity*: How representative is the experimental situation of the relevant situations outside the laboratory?

(l) *External Validity*: How reproducible is the experimental effect in relevant situations outside the laboratory?

Readers familiar with Ian Hacking's (1983) primer on the epistemology of experiment can keep this distinction straight by thinking of (k) as treating experiment as a method that generates results for *representing* reality, whereas (l) treats it as a method that generates results for *intervening* in reality. Thus, the goal of (k) is to use the lab as a means of understanding some part of the world as it exists independently of what happens in the lab, while the goal of (l) is to use the lab as a means of making some part of the world conform to what happens in the lab.

Since Brown evaluates the experiments against the historical record, he is therefore probably best read as primarily lodging an (k)-type validity challenge. And it would seem that, in terms of ecological validity, Brown's point is unassailable. The task of undergraduates trying to discover the rule governing a number series is nothing like the problem that scientists face when they try to make sense of some laboratory phenomenon. The intuitive pull of this claim may appear so strong, given the obvious surface differences between the two situations, that we do not need any more proof to be convinced that Brown is right. However, to be convinced on the basis of that intuition alone would be tantamount to arguing, as a seventeenth century scholastic might, that balls rolling down an inclined plane can tell us nothing about the nature of celestial motion because the experimental setting is too simplified, as well as too far removed, both perceptually and conceptually, to be an adequate model of the naturally occurring phenomenon (cf. Houts & Gholson 1989). The historical antidote for the scholastic's way of thinking was Galileo's introduction of a strong primary–secondary quality distinction, which showed that there is literally more to an experimental effect representing reality than meets the eye (Koyre 1978).

I do not expect that appealing to a historical analogy will settle the issue of ecological validity, but I hope that it will delay our Aristotelian reflexes long enough for the proper ecological validity tests to be done. An important test would be to examine whether the modeled population—in this case, scientists—find the tasks given to subjects in the experiment relevantly similar to the tasks in which they are normally engaged. Again, it might seem that the target population would immediately fail to see any similarity. Here questionnaire design becomes important, since the target scientists must be asked to make focused remarks about particular reasoning skills involved in the experiment, rather than simply offer their general impressions, which would no doubt incorporate an awareness of the obvious dissimilarities between the undergraduates' situation and their own. Needless to say, there is the potential for scientists' misleadingly describing their experience by falling back on stock ways of talking, perhaps even derived from positivist philosophy of science. As in the case of the cognitivist history of science, protocol analysis (Ericsson & Simon 1984) can prove a useful diagnostic tool here.

But none of what I have said so far denies that, at the end of the day, Brown will be proven right in his Aristotelian intuition. However, even if Brown's (k)-type validity challenge holds up, it is the (l)-type validity of the experiments that is of greater concern to the normatively oriented philosopher of science. To illustrate the difference implied here, consider a series of experiments that Michael Gorman and his associates ran, following up on earlier research that showed that both scientists and non-scientists approached problem-solving with a strong confirmation bias (Gorman et al. 1984, 1987). The subjects' task in these experiments is (yet again) to discover the rule governing a number series by proposing other number series as hypotheses and then being told whether or not the test series conform to the rule. Gorman observed that in the original experiments (Tweney et al. 1981, part 4), the subjects generally worked alone and were not informed of strategies for discovering the rule. Working on the assumption that the discovery process would be facilitated by subjects' working cooperatively in groups and by being instructed in a falsification strategy, Gorman gave the subjects the same task under these revised conditions and found that their problem-solving effectiveness improved dramatically.

Arguably, Gorman's experiments are even less ecologically valid than the original ones, since scientists do not typically face a problem prepackaged with instructions on how to solve it. Moreover, unlike some of the original experiments, Gorman's never involved scientists as subjects. But even granting that Gorman's experiments are not representative of the conditions of scientific reasoning as it normally occurs, it still does not follow that his results could not be reproduced in normal settings. I am assuming, for the sake of argument, the desirability of having Gorman's results writ large, as well as the feasibility of introducing Gorman's social and pedagogical dimensions into actual research settings. Admittedly, these are large, but not unreasonable, assumptions, to which much thought has been given by experimental social psychologists (e.g., Campbell & Stanley 1963, Wuebben et al. 1974). If it comes to pass that Gorman's results hold up in a variety of research settings, then a falsificationist group problem-solving strategy may become part of ordinary scientific training. Over time, that would serve to change what scientists did normally, so that, in the long term, Gorman's experiments would *become* ecologically valid.

There are three points to make about this argument.

First, the difference between (k)- and (l)-type validity gets complicated, once the psychologist shifts her interest from reliably improving scientists' problem-solving performance to understanding what it is about the scientists and their natural surroundings that enabled such an improvement to take place. In other words, once we move from seeking regularities to seeking their underlying causes, ecological validity needs to be given a more dynamic interpretation. Up to this point, it has looked as though, insofar as they were pursuing ecological validity, the psychologists were simply trying to model what real scientists actually do, rather than what the scientists have the potential for doing. However, the concept of ecological validity can easily be extended to cover attempts to represent the scientists' real potential for change, which presumably they have independently of what the psychologists discover about undergraduates in their labs. This extended sense of ecological validity is a prerequisite for the psychologists' experiments having external validity, since it is clear that an effect cannot be reproduced in an environment outside the lab (say, among real scientists) unless the environment already contains the potential for having the effect brought about (cf. Bhaskar 1979).

The second point is that in this discussion, and in much of what follows, I presume that philosophers would want to intervene in scientific practice in order to improve problem-solving effectiveness and efficiency; hence, the repeated use of Popper's falsificationism as an example. However, I do this only as a matter of convention, though a convention that involves more than a token gesture to internalism. Brown is correct to point out the affinities between the aspects of scientific reasoning that interest experimental cognitive psychologists and those that interest philosophers wedded to the internal history of science. Roughly speaking, both are looking for problem-solving strategies that work the most often for the most different cases. Brown is incorrect, however, in thinking that these affinities amount to guilt by association. On the contrary, given the inertialist picture of rationality presupposed by the internal history of science, it would seem that in only a setting as controlled as a laboratory experiment could one properly test hypotheses about the propensities of the autonomous scientific reasoner. Realization of this point has led Shadish and Neimeyer (1989) to argue that whatever difficulties arise in reproducing psychology of science results outside the laboratory constitute indirect evidence for the ineliminability of an externalist, context-bound understanding of scientific reasoning.

But having said that, I think that, ultimately, the most interesting interventions will involve disciplining the flow of scientific communication. There are many independent variables to consider here, the most important being these: the number of scientists and their distance from one another in time and space, the ease of access that they have to each other's work, the time that scientists distribute between accessing the work of others and doing their own, the size and structure of a journal article or other unit of scientific communication, the format for making reference to the work of others (how many citations, of what sort, for what reason), linguistic consistency across scientists, and constraints on the introduction and discussion of topics. My hunch is that the epistemological differences normally associated with the difference between, say, physics and literary criticism are not due to differences in subject matter or even in the techniques of inquiry (e.g., apparatus manipulation vs. document reading), but rather to differences in the variables just mentioned. If so, then manipulating these variables on a large scale could render literary criticism as "rigorous" as physics, or physics as "fluid" as literary criticism. I imagine that if this point were to be experimentally demonstrated, then ("axioaetiotically" speaking) the relative value that we place on the various branches of knowledge would be drastically altered. But this conjecture must await another opportunity for formal refutation (cf. Fuller 1988b, ch. 8, on "disciplinary ambivalence").

Finally, if we think of the "normative" dimension of the philosophy of science as being concerned with interventions designed to improve the production of knowledge, then the philosopher is finally doing something that a literary critic is not typically doing when she appraises the quality of a work. The focus of the normative project has thus shifted from *science evaluation* to *science policy*. This is an important point, given that we have seen that internalist history of science has unwittingly promoted a more diminished sense of the normative. Moreover, some "postmodernist" philosophers (e.g., Rorty 1979, part 3) have openly embraced the model of philosopher of science as literary critic, with the relevant literature being scientific texts. However, the price that we must pay for cultivating a more robust normative sensibility is that determining the course of science becomes a thoroughly *political* problem. In what follows, in order to highlight the epistemologically interesting issues involved in politicizing science, I reconstruct the Neo-Kantian debate over the status of economics as science and policy.

9. The Perils and Possibilities of Modeling Norms: Some Lessons from the History of Economics

At the turn of the century, the most philosophically active discussions of the role of modeling in science occurred in the course of debates over concept formation in the social sciences. These discussions took as their point of departure John Stuart Mill's model of the rational economic agent, *Homo oeconomicus*, the object of inquiry in political economy and subsequently of neoclassical economics. Such an agent was defined as someone who consistently preferred greater (to lesser) wealth, less (to more) work, more (to less) immediate consumption. The goal of the science of political economy, then, was to determine the laws that govern the activities of such an agent under a variety of circumstances (Mill 1843, vol. 2).

The German Neo-Kantian philosophers were the ones largely interested in understanding the sense in which Mill and his successors were "modeling" economic phenomena. They structured their debates around specifying the cognitive process by which the model was abstracted: Was it by *isolating* certain features of real economic agents that economists believed, for some reason or other, could be studied as a closed system of interacting variables? Or, was it by *generalizing* on the features of those agents that recurred most frequently in the real life situations of most interest to the economist? The former process produced what was generally known as *strict-types*, which Max Weber would later immortalize as "ideal types," whereas the latter process produced *real types* (Machlup 1978, ch. 9). What follows is a rational reconstruction of this debate, generally called the *Methodenstreit* in political economy, with an eye to its implications for a scientific social policy, a special case of normative naturalism (cf. Proctor 1991, part 2; Bryant 1985, ch. 3; Manicas 1986, ch. 7). Laying out the field of play will allow us then to focus on the part of society for which we would like to develop a scientific approach to policy, namely, science itself. Aside from this self-serving aim, the general significance of this debate cannot be underestimated, as its aftermath engendered the disciplinary fissure between neoclassical economics (the winners, who were strict-typists) and political science (the losers, who were real-typists).

The strict-type view followed up, to a large extent, the classical empiricist and idealist account of abstraction, namely, as the process of distortion or simplification by which finite minds managed the infinite complexity of reality. This view, which ultimately descends from Aristotle's *aphairesis*, was often associated with the problem of conveying in words the inexhaustible richness of intuition. The key philosophical defender of this view at the turn of the century was Wilhelm Dilthey, with Karl Menger numbered among its champions in the social sciences. It may appear strange to couple Dilthey, the great hermeneutical methodologist, with Menger, the original theorist of marginal utility whose work undergirds contemporary neoclassical economics. After all, the practice of deep textual interpretation would seem to be the very antithesis of constructing supply and demand curves. However, the natural complementarity of the two tasks becomes especially clear from Menger's methodological writings (cf. Manicas 1986, p. 134, for a similar point about Max Weber).

Unlike more positivistically oriented thinkers, as Mill himself was taken to be (Brown 1986, ch. 8), Menger did not take the abstracted features of *Homo oeconomicus* as a position of cognitive strength from which one could predict and control the behavior of real agents, as one could of real bodies via the abstractions of classical mechanics. Rather, Menger understood the strict-type to be a position of cognitive weakness to which the economist had to retreat because she could not, in principle, disentangle the complex consequences of many real agents interacting at once over long periods of time in the marketplace. Of course, Menger did not deny that the economist could predict the behavior of any finite number of abstract economic agents (i.e., ideal utility maximizers) whose propensities to produce, save, and consume had been suitably quantified. But whether such hypothetical interaction could be then shown to model the essential character of any real economic environment—past, present, or future—was, strictly speaking, beyond the ken of economic science.

This is not to say that no conclusions whatsoever could be drawn from the model to the real case, but simply that the pattern of inference would not be uniform from case to case. The pattern would depend most of all on the cultural traditions that form the backdrop of the particular economic transaction under study. This, in turn, would involve a largely intuition-based inquiry into the self-understandings of the real agents. At this point, we can see the hermeneutical hand that has been guiding this line of reasoning all along. It was common enough to methodological debates at this time (e.g., a version of it can be found in Wundt's strictures on the role of experimental psychology in understanding higher mental processes [Fuller 1983]) and appears most clearly today in the "hermeneutical economism" of Friedrich von Hayek (1985) and a fellow Viennese economist who increasingly gravitated toward phenomenology, Alfred Schutz (Prendergast 1986).

The issue becomes somewhat complicated for the strict-type at this point, since there are at least two reasons why a strict type may fail to have any predictive value. These two reasons pertain, respectively, to what I called, in the previous section, "ecological" and "external" validity. For example, it may be that, say, Mill has indeed isolated the essential features of the consumption and production patterns of *Homo oeconomicus*, but that these features are, under normal circumstances, altered by local considerations that vary on a case to case basis. On this view, no one truly acts as a rational economic agent because no one is ever in a situation that calls *only* for economic deliberations. But the features isolated by Mill are no less real; they are just routinely suppressed. A rather different reason for the predictive failure of a strict type may be that the isolated features have only prescriptive force. In other words, Mill may be taken, not to have discovered the essence of *Homo oeconomicus*, but rather to have contributed to a strategy for converting real economic agents to rational ones by characterizing the sorts of features that they would have to maximize. In that case, there is no need to decide whether the kind of self-discipline (or social control) required of real agents to become rational serves to elicit or to transform natural human tendencies—just as long as the agent can, by some means or other, be eventually made into a rational economic being.

As for the real type view of scientific models, its lineage can be traced to Aristotle's *korismos*, the extraction of what is common to a set of particulars: in short, the abstraction to universals. Following Mill, this process was understood as a species of induction from many cases, with the number of abstractable universals given by the natural propensities of the cases themselves. Whereas the price of abstraction in strict typing was the loss of society's natural complexity, the price in real typing is the loss of the individual's natural uniqueness. However, advocates of the real type view, such as the philosopher Heinrich

Rickert and the political economist Gustav Schmoller, saw the price of abstraction as ultimately yielding a net cognitive gain, in that it would allow for the identification of statistical regularities, that could, in turn, form the basis of rational economic policy. Max Weber's notorious skepticism over the possibility of such policy—at least of a policy that could be inductively inferred from statistical indicators—rested, in large part, on his refusal to grant that abstraction could be sufficiently passive to permit so-called natural propensities in economic phenomena to impose themselves on the receptive economist's mind. In the Weberian lingo, the "value-neutrality" of economic knowledge to policy is a direct consequence of the "value-relevance" that informs the gathering of economic knowledge. Self-interested inquirers probably make for self-deceived policymakers.

This last remark raises an important point about the distinction between the isolating and the generalizing approaches to modeling the economic agent, namely, the complementary character of the two approaches. Strict-typists, like Menger and Weber, portray the modeling process as requiring the conceptual—and perhaps experimental—intervention of the economist, since economic reality does not naturally reveal regularities to the passive spectator. Not surprisingly, once a closed system of interacting variables has been isolated, it still does not follow that the system can be mechanically applied (i.e., without additional intervention) to determine the course of economic reality. As a result, the specter of a continually distorted economic reality raised by the strict-typist's interventionist epistemology leads her to a non-interventionist (laissez-faire, in economists' terms) policy orientation. By contrast, for the real-typist like Schmoller, the question is not whether to intervene actively or to observe passively, but rather whether to focus the economist's gaze on the overall picture of economic activity or simply on its details. No intervention would be involved in either case, as the difference between expanding and contracting the focal range amounts to manipulating one's observation of the phenomena—not the phenomena themselves. Armed with this non-interventionist epistemology, the real-typist has no conceptual difficulties with intervening to determine the future of the economy, since the intervention would simply make already existing tendencies more evident.

For a sense of the practical difference made by this shift in, so to speak, *the burden of intervention* from epistemology to policy, consider the alternative analyses that Menger and Schmoller would give to the impact that the announcement of state economic policy has on the future of the economy. Menger would treat the announcement as yet another factor whose effects will be differentially distributed throughout the economy, given the variety of reactions to be expected from the economic agents who learn of it. He would be especially concerned about the possible self-defeating qualities of the announcement, as each agent tries to turn it to her own individual advantage. Thus, Menger's analysis would resist seeing the policy statement as a causally inert description of what is likely to happen, emphasizing instead its (perhaps unintentionally) interventionist character and hence its tendency to confound the course of economic events.

For his part, Schmoller would regard the announcement as instructing economic agents to prepare for, or maybe hasten, a state of affairs that is already objectively probable. In addition, the announce-ment provides convenient language for defining subsequent events as either facilitating or impeding the policy's realization. For example, if state policy calls for the expansion of certain sectors of the economy and this fails to materialize, then the state can explain the failure in terms of a local irregularity of one sort or another. In any case, on Schmoller's analysis, state economic policy brings into focus—in a way that only a central authority can—certain overarching economic trends to which the economic agents themselves are only dimly aware of contributing. Greater awareness of these trends is expected only to enhance the collective rationality of the agents' practice and thereby bring the trends into sharper relief.

A good way of epitomizing the discussion up to this point is by listing the various ways in which experimental intervention might be used to study norm-governed behavior in science. I shall draw on Kuhn's (1977, ch. 3) account of the changing role of experimentation in the history of science since the the Middle Ages. Kuhn proposes at least four roles that experiment has played, two of which may be associated with a real type approach and two with a strict type approach. Of course, in the cases Kuhn himself considers, the aim of experiment is to elicit physical laws and other regularities typically studied by the natural sciences. For our own purposes, the relevant regularities governing the behavior of scientists are better characterized as "methods" than as "laws." And for the sake of simplicity, I will repeatedly refer to Popper's falsificationism—the methodological strategy whereby the scientist is oriented toward eliminating alterna-tive hypotheses for explaining a given phenomenon—as the relevant regularity under experimental investigation.

1. Falsificationism is studied experimentally by testing for the extent to which it causes desirable scientific behavior. This strategy is attuned to the interests of the strict-typist. The experimenter here need only presume that the norm can be discerned in a significant amount of desirable scientific behavior, but she need not presume

that it is the norm itself—rather than something regularly correlated with the norm—that causes the desirable behavior. The experimenter would then proceed to isolate these covarying features of what the scientists do to see whether they alone, without the intent to falsify, would have accounted for most of the desirable behavior. For example, it may be that research projects designed to whittle down the number of competing hypotheses in a field are also the ones that are most highly funded, and not surprisingly they most often produce scientifically desirable outcomes. Moreover, it may even be that falsificationist projects are most amply funded because the decision-makers at the funding agencies (persuaded by Popperian ideology, no doubt) believe that such projects are methodologically most astute. Yet these conditions would be entirely compatible with the experimental finding that it is simply the generous level of funding that has brought about desirable scientific outcomes, and that were research projects oriented by other methodologies so funded, they too would be just as likely to succeed.

2. The real-typist experimentally intervenes to apply the norm to new domains. If falsificationism is presumed to be a successful strategy among natural scientists, then the experimenter would see whether the norm produces similarly desirable results when social scientists are placed under its strictures. It may turn out, however, that deploying falsificationism in the relatively primitive conceptual environment of the social sciences eliminates all the available hypotheses before they have had a fair chance to be empirically elaborated. (Indeed, this was one of Lakatos' main objections to strict falsificationism.) But notice that although such a result may bode ill for either the norm or the domain to which it has been applied, the negative result does not detract from the measure of success that the norm enjoyed in its original domain.

3. Falsificationism is studied experimentally by removing empirical obstacles to its appearance, such as by offering incentives for scientists' making and meeting criticisms. Keeping with the strict-typist's view of things, this involves presuming that the norm does not normally govern realistic scientific activity, but that it prima facie ought to. The experimenter's task, then, is to manufacture the conditions for revealing falsification in action and afterwards to decide whether these conditions can be approximated at an affordable expense in more realistic settings. The "naturalness" or "artificiality" of falsificationism as a scientific norm is therefore a function of the cost of approximating the ideal conditions on a regular basis vis-à-vis the benefit that is likely to result from such routinized effort.

4. More in keeping with a real-typist's sensibilities is an

experiment that simply involves increasing the level of regularity with which the norm is exhibited in scientists' behavior. The presumption here is that scientists already successfully falsify hypotheses under a wide variety of conditions, and that the remaining deviations from consistent falsificationism may be corrected by adjusting some locally occurring factors that interfere with the scientists' motivational structure. Skinner (1954) would call this process "operant reinforcement." However, as we will see below, there is actually more to the issue of exhibiting the norm than either the cost–benefit analysis of the third strategy or the local adjustments of this last one would suggest.

10. The Big Problem: How to Take the First Step Toward Improving Science?

I have been belaboring the nuances of the Neo-Kantian debate over concept formation in the social sciences in order to provide a vantage point from which to reconstruct recent philosophy of science debates over the compatibility of the natural and the normative. I began Chapter Three by bemoaning Laudan's restriction of the normative to the value judgments of today's historian of science. Yet, as we saw at the start of this chapter, Laudan's retreat seems to reflect an implicit chain of command in which philosophers of science stand between those from whom they ought to take advice and those to whom they ought to give it—scientists making up the former category and historians of science the latter. Normally, the chain of command is explained in terms of the problem of induction which blocks any uniform pattern of inference from what worked for science in the past to what is likely to work in the future. Nevertheless, so it is then said, the philosopher may at least have something interesting to say about what has worked for science in the past.

However, this explanation, which trades heavily on the philosopher's ignorance of the future, is often blended with another account, one that emphasizes that the philosopher is only one among many whose actions are constitutive of what science will become: To wit, the future is open with regard to the sorts of values that may be promoted by the pursuit of science, and the values we select depend on the sort of society that we generally wish to bring about. Clearly, the actual indeterminacy of the future imposes a more trenchant limitation on the philosopher's normative powers than mere ignorance, especially as it suggests that our epistemic track record may have little bearing on what we ought to do, should we decide to pursue a social order devoted to values that are radically different from those in which science has flourished in the past. At this point, the prospect of drawing normative conclusions from the history of science seems most daunting, which perhaps explains the reluctance of philosophers to cast their gaze into the future.

Still, the openness of the future is hardly the biggest obstacle to the philosopher of science trying to satisfy her normative urges. For even in this case, history can serve as a fairly straightforward guide. After all, to say that the science of tomorrow might fail to resemble any earlier science is not to say that the values promoted by such science had never been promoted before. Perhaps the radicalness will lie (as Auguste Comte, for one, thought it would) in the use of scientific institutions to promote values previously promoted by religion. Let us say that we want tomorrow's science to make people more charitable in their treatment of others. Although there is no precedent for science functioning in this capacity, the historian can still provide a sense of how this might work by examining (i) the forms of social organization that tend to promote charity, (ii) the differences between those forms and the forms that have tended to promote modern science, and (iii) the extent to which the fruits of modern science depend on its being pursued within forms that are less likely to promote charity. The last task especially will require an extra dose of Mill's canons of induction and counterfactual history. But as long as the target values are defined in terms of a demonstrable difference in human behavior that their promotion would make, then the identity of tomorrow's science is empirically decidable (cf. Hirsch 1976, ch. 6, on the empirical character of literary evaluation).

At the risk of sounding crassly positivistic, I submit that the above strategy is "easier said than done" mainly because contemporary discourse on values is distinguished by words such as "charity" that denote a range of possible states of affairs, many of which are mutually incompatible—at least practically and perhaps even logically. This basic unclarity in the key value concepts makes it difficult for axiologists to think in terms of concrete outcomes, time frames, and constraints on acceptable side effects. However, if axiologists were to redescribe the desired values in terms of their policy implications, then the issue would be no less decidable than any other empirical one (cf. Fuller 1988b, ch. 12). The *Methodenstreit* is itself one of the most important cases in which differences in the key value concepts remained unresolved for the duration of the debate. Since the strict-typists won the scientific high road that was at stake, we tend nowadays to conceive of the relation between facts and values in Weberian terms, namely, that values are trenchantly subjective, indeed of such personal importance that they should not be treated as technical questions of fact. Presupposed in this line of reasoning is that it would be wrong to treat value questions as matters of fact because that would be unduly discouraging. The empirical world cannot support the fullest realization of everyone's values, which means that tradeoffs must be made. The social scientist can help by laying out the possible trade-offs. But in the end, Weber is a Kantian: That is, value pursuits are ends in themselves, regardless of their empirical feasibility.

However, had Schmoller and the real-typists won the *Methoden-streit*, we would be living with a different fact–value sensibility. Schmoller supposed that values were the objective products of a shared cultural history, as epitomized by one's national identity. Thus, basic values were unlikely to vary as much from citizen to citizen as Weber supposed. Moreover, these basic values enable citizens to identify sympathetically with one another's more particular interests, which in turn encourages everyone to put in whatever effort it takes to generate the resources needed to satisfy the ends of their fellow citizens. Not surprisingly, then, Weber dubbed Schmoller's party "socialists of the chair." Although Weber successfully portrayed Schmoller et al. as naive utopians, one is led to wonder who really has the more mystified view of what follows from, so to speak, *value scarcity*, that is, the world not being able to support the realization of everyone's values at once: Schmoller, who treats it as a technical issue to be resolved by collectively producing more of the relevant goods; or Weber, who treats it as a brute fact to which people adapt by pursuing their ends as a matter of principle, without regard to consequences.

The problem of value scarcity moves us closer to the most important obstacle facing a normative philosophy of science. It is an obstacle that would remain even after closure had been reached on what the future ought to look like. For even if (by either isolating or generalizing features) we knew which qualities of the scientific reasoner ought to be maximized in order to promote the growth of science, and even if maximizing those qualities were compatible with other qualities that we deem desirable in the citizens of tomorrow's society, those two facts alone would be insufficient for addressing the crucial question of what exactly ought we do to bring about this socially acceptable science of the future.

Any doubts that you may have had about the epistemic soundness of abstracting real-types and turning them into the basis of social policy should now be surfacing. For all their avowed inductivism, real-typists like Schmoller share a key assumption with Descartes and most other philosophers who have seen the discovery of the True Method as a source of instant enlightenment and, hence, worthy of immediate publicity: namely, that once revealed to the relevant audience, the

Method will be seen for what it is, leading the audience to do whatever they can to conform to its strictures. In short, real-typists are dedicated to the *self-certifying* and *self-edifying* character of the Method. Thus, notwithstanding their avowed policy interests, real-typists leave no room for a *politics* to mediate between the scholarly reflection that informed the philosopher's discovery of the Method and the behavior that will demonstrate that the Method has been successfully followed by the audience. It is precisely this tertium quid, occupied by things political, to which strict-typists like Menger and Weber are especially sensitive. Indeed, it leads them to question the possibility of scientific social policy. For within this scantily charted realm between normative theory and actual practice lie the philosophically impeachable pursuits of persuasion, manipulation, and coercion, which, for better or worse, form the repertoire of techniques for "applying" knowledge.

A few philosophers of science have already seen the relevance of this last point for their studies. The work of Edward Manier (1986) is especially noteworthy for showing that what, at the "conceptual" level, may appear to be two sciences ripe for synthesis—behavioral psychology and neurophysiology—pose major problems once a practitioner of one of these sciences attempts to apply her concepts to the data gathered by practitioners of the other. Such efforts at cross-disciplinary negotiation nearly always fail, and if they do not, then the practitioners run the risk of becoming marginalized by their respective disciplines and forced to start a new discipline of their own. Manier has argued that philosophers systematically underestimate the difficulties of integrating bodies of knowledge by thinking of the issue purely in terms of constructing a Quinean translation manual between sentences in the two sciences. Even assuming (as positivist folk wisdom would have it) that knowledge production is enhanced by synthesis, the fact that the philosopher can regiment the discourses of psychology and physiology so as to draw the appropriate terminological correspondences between mental states and physical states is still insufficient to show that the scientists would themselves find such a translation mutually agreeable.

This is where politics enters. Practitioners of distinct disciplines are justifiably unmoved by philosophical appeals to a "common referent" to which a sentence in each discipline corresponds, until the argument has been filtered through the appropriate disciplinary rhetorics—an especially difficult feat when the argument must filter through two such rhetorics simultaneously. Moreover, the philosophical elixir for virtually every intertheoretic incommensurability—namely, a "topic-neutral" metalanguage—will likely strike both psychologists and physiologists as an open admission that their differences cannot be resolved *on their own terms*. Indeed, the very idea of resorting to a metalanguage misses the point of interdisciplinary negotiation, which is not to discover a discourse sufficiently comprehensive to render the two disciplines redundant, but rather to construct a pidgin or creole dialect to cover the domains where the concerns of the two disciplines overlap (cf. Fuller 1987b, p. 358; Aitchison 1981, ch. 12).

More generally, my point is this: Although much effort has been invested in determining whether scientists have historically adhered to (say) a falsificationist methodology, whether scientists are cognitively well-disposed to such adherence, and whether falsificationism is likely to promote good science, those who have investigated these matters have generally assumed that if the questions are answered affirmatively, then training tomorrow's scientists to falsify hypotheses is an assignment of no great political import, in that it "simply" involves sitting the scientists in a classroom and informing them appropriately.

What is missing here is any awareness of what has been responsible for falsification's epistemic success. For insofar as falsificationism has been a successful methodological strategy, it has always been pursued, not as an isolated end, but in conjunction with an array of other scientific and non-scientific interests, and at varying degrees of self-consciousness. Were falsificationism now raised to an explicit policy goal, a veritable "indicator" of desirable science, a whole host of new problems would arise. I believe that most of these would be the result of the attempts of scientists to economize in their efforts to convince the relevant cognitive authority that their research adheres to the falsificationist strictures. For example, a style of reporting research may be perfected that leads the reader to presume (rightly or not) that the reported experiment was conducted in a methodologically sound manner. Of course, the net result of countermoves of this sort would be to undermine the intent of the policy.

In short, it may be that part of the original success of falsificationism was due to scientists' being ignorant of the methodology's crucial role in their pursuits. As a result, the scientists had no particular incentive to publicize their conformity to falsificationist strictures beyond what they normally—and truly—did. But my making the point this baldly may, in turn, give the misleading impression that at issue here is a weakness of will on the part of scientists: specifically, an inability to curb their achievement drive, which moves them to cut procedural corners whenever it seems that the likelihood of being caught is low. Although I have said nothing so far to gainsay this interpretation and indeed it is the sort of interpretation that would be expected of a positivist or a Popperian, I would rather opt for another line of thinking, one that calls into question the *cognitive* character of normative constraints presupposed in the weakness-of-will analysis.

11. Behaviorally Speaking, the Options Are Numerous but Disparate

When we looked earlier at the various ways in which experiments can be used to determine and extend normative scientific activity, we found that the success of such experiments was generally measured by whether scientists behaved in greater conformity with the norm. Yet, sometimes it seemed to matter whether the scientists were disposed to avow the norm when asked. Here, then, are two of at least four distinct ways of measuring normativity—or from a more behavioral standpoint, displaying our knowledge of norms. Once we spell out the four potential indicators, it will become clear that there is no prima facie reason for believing that enhancing the display of one indicator of normative activity automatically serves to enhance the others. Indeed, the management of these *divergent* indicators—each of which is a proper measure of normativity in some important settings—is the principal source of problems in the application of norms.

My discussion of scientific norms draws on Peter Collett's (1977) research into rule governance. For our purposes, the first telling point that Collett, a social psychologist, makes about the nature of norms is that if a community is said to be governed by a unified set of norms, then it must be possible to state the norms without distinguishing between the agent's conforming to the norms and her recognizing someone else's behavior as conforming to those norms. There are several reasons for imposing this condition, some pertaining to a sense of equal justice for judges and judged, and some to a sense that playing the judge makes one better at playing the judged (and vice versa), which suggests that the two roles share a common knowledge base. For example, in Chomsky's generative grammar, which is supposed to be the mind's repository of linguistic norms, the same rules of syntax are said to govern both the interpretation and the production of sentences. The difference, then, between a hearer and a speaker of the same language is merely the physiological process that embodies their common grammar.

I suspect that a similar assumption is made by philosophers of science, which would explain why they have been singularly indifferent to the question of whether a norm like falsificationism is supposed to govern the way one *conducts* research or the way one *evaluates* it. Yet, the reason why heightening the sensitivity of scientists to the signs of falsifiability is unlikely to get them to do more falsifiable research is precisely that doing research and evaluating research are not symmetrical activities governed by the same norms. In other words, the philosopher's pedagogical efforts may have backfired because she tried to get the scientists to *conduct* falsifiable research by teaching them how to *evaluate* research for its falsifiability. If the philosopher had not been inclined, at an abstract level, to presume the symmetry of doing and evaluating research, she would have little motivation in the particular case for making such a presumption. The difference, after all, is between the writer and the editor of journal articles—two roles that a scientist may play equally well, but not necessarily by developing skills that transfer from one role to the other.

The picture of the scientist's psychology that I am painting deviates significantly from that of the coherent reasoner, all of whose practices—however outwardly different they may appear—are informed by the same normative principles (at least under ideal conditions). In its place, I am proposing a much more internally fragmented view of the scientist. My source is Jerry Fodor's (1983) modularity thesis, but with a behaviorist twist. That is, for each way of displaying normative behavior, there is an "input module" that registers information that is relatively inaccessible, or "impenetrable," to the other modules and any central cognitive processing unit that might be designed to integrate the information received by the modules. (The subjunctive mood of the "might" gives away my doubts that such a unit is to be found in the individual scientist: cf. Fodor 1983, pp. 107ff., Kornblith 1987.)

However, I depart from Fodor in resisting the Cartesian connection between the cognitive biases of the modules being *innate* and their being *unchangeable*. Rather, I follow behaviorists like Skinner (1954) in interpreting these biases as operants that are selectively reinforced by the environments in which they are exercised. Take Fodor's own prime example of the cognitive impenetrability of input modules, the Mueller-Lyer illusion, in which we can have propositional knowledge of how the illusion works, yet that knowledge does not seem to stop us from continuing to see it in its illusory state. What is impenetrable to what here? Fodor seems to think that the appropriate level of description is fairly specific, that is, that there is an a priori fixed visual bias to seeing the Mueller-Lyer illusion, which hinders our incorporating the dis-illusioning bits of perceptual psychology. By contrast, I would locate the bias at a more general level, namely, that we are not cognitively equipped with any systematic way of integrating visual and discursive information. Yet, the particular biases that the modules display at a given point (i.e., the sorts of errors we tend to make) are best explained in terms of selective reinforcement.

Thus, if I come to see the Mueller-Lyer illusion for what it is, it will not be because my having heard or read about perceptual psychology precipitated a gestalt switch in my perception of the figure. At least, that could not be the whole story, since there is no "natural" way to translate speech into sight. In that respect, Fodor is an effective antidote to the view associated with Hanson (1958) and Kuhn (1970a) concerning the "theory-ladenness" of observation. However, contra Fodor, if my vision module is exposed to the appropriate cues in the environment on a regular basis, then I can unlearn the perceptual bias that the illusion embodies. The important point here is that these cues may emerge quite unintentionally, as I do things one of whose by-products is that the world appears different to me (e.g., I add a new artifact to the landscape that triggers a reorganization of my perceptual horizon; cf. Heelan 1983). If theories (as linguistic entities) are involved at all in this process, they enter probably after the fact and necessarily by convention, since, on the modular view, a change in what I see can be caused only by stimulating the visual modality. Apparently, this position is close to one that Popper himself reached while training in educational psychology under Karl Buehler, largely on the basis of watching children as they were taught to draw (Bartley 1974).

To suggest, as I do, that science is governed, not by a unified set of norms, but by sets of role-encapsulated norms is to portray knowledge production, even in its ideal state, as socially fragmented and perhaps internally divided (cf. Minsky 1986). I must admit guilt to this charge and the implications that follow from it. Among the most interesting of these is that the very identity of a scientist's normative structure may vary, depending on the behavioral display that is taken to be the canonical indicator of her acting in a properly disciplined manner. Should the observer take the norms implicitly governing the scientist's research practice as the basis on which to evaluate her avowed theory of research, or vice versa? In fact, the matter is considerably more complicated, given that Collett has identified *four* separate normative display modalities. No doubt, much of the difficulty that philosophers of science continue to have about characterizing epistemic norms—and who should be taken as (and as not) conforming to those norms—stems from philosophers' not being alive to the possibility that the different behavior displays may be best explained as being governed by differernt norms. These difficulties would thus be resolved if philosophers took the modularity thesis more seriously and did not think of norms on the Chomsky model of one abstract grammar with multiple behavioral instantiations. But again, to make things easy, I will describe each of Collett's four normative displays as being displays of the same norm, falsificationism.

Collett observes that there are four ways of determining the rules governing a community. Let us say that a philosopher of science, or some anthropologist in her employ, has determined that a given scientific community is governed by falsificationist norms. How might he have determined this fact? One obvious way would simply be to ask the scientists to state the norms governing their professional behavior. No doubt, each scientist's list would be somewhat different, but among the overlapping statements would be one of the falsification principle. Another way of determining the norms would be to observe, "at a distance," the sorts of things scientists in the community normally do, with the aim of arriving at the most economical expression of that regularity. You will recall this approach as underlying the real-typist's strategy for getting at the essence of *Homo oeconomicus*. A third way, popular among linguists (Greene 1972) and logical positivists (Carnap 1956) as well as their analytic offspring, is to survey the scientists' intuitions of normativity by asking them whether certain hypothetical cases of research activity fall inside or outside the normative bounds and then inferring the norm implicitly governing these intuitive judgments. Finally, the philosopher may determine the scientists to be falsifiers in the course of her trying to account for the occasions in which they actually reject, criticize, or otherwise impose sanctions on the work of their colleagues, as these would be moments when the scientists instinctively recognize behavior as falling outside the standards of acceptability (cf. Simmel 1950).

Notice that the first two ways of displaying normativity pertain to a scientist's own activities, whereas the last two pertain to how that scientist judges the activities of others, whether they be hypothetical or real. Moreover, the first and third ways require that the philosopher intrude upon the scientist's ordinary activities, asking her to make judgments in the abstract. By contrast, the scientist would be better not apprised of the philosopher's presence in cases two and four, since here the philosopher is interested in a certain "natural" sense of normativity. However, the most important difference in the four forms of falsification emerges when we consider some likely stories by which the scientists *learned* to display each form and the environmental conditions in which each display is *reinforced*.

For example, the scientists probably learned to state the falsification principle by taking a methodology course or reading a Popperian tract that promised to situate their research in a larger epistemic context. In either case, the learning experience was rather removed from their day-to-day activities as scientists. However, the regularity with which scientists behave in conformity with falsificationism probably did begin with their "hands-on" training at university and has since been periodically reinforced by rewards both internal and external to the scientific process. As for their intuitive judgments, scientists probably have received no special training, which makes them highly dependent on the actual wording of the hypothetical cases and the ease with which these cases can be represented in terms of situations drawn from their professional experience. If anything, in the course of being questioned by the philosopher, the scientists will learn how to evaluate these cases (cf. Webb et al. 1969, p. 22). Finally, learning when to reject and criticize a piece of work is perhaps the most difficult display of normative behavior to explain. However, since it may well also be the most important in the entire epistemic process, it merits a more detailed treatment.

Given traditional philosophical images of the scientist, it should come as no surprise that the role played by negative sanctions in maintaining the social order of science has been understated. But I suspect that this is only because philosophers have a fairly simpleminded view of what these negative sanctions amount to. On the one hand, we tend to imagine a Hobbesian sovereign who plays on the subject's aversion to pain as a means of getting her to act in some desired fashion. On the other hand, scientists are normally portrayed as dwelling in an environment that promotes intellectual diversity and innovation, an image that resists the sort of conformity that negative sanctions would seem to impose. But, as I have suggested, the contrast is all too black and white (cf. Goodin 1980).

In the first place, this simpleminded view highlights the *coercive* character of negative sanctions at the expense of their *manipulative* character. The distinction between coercion and manipulation that I have in mind corresponds to a Machiavellian sense of the difference between an antipolitical and a truly political approach to government. In fact, were Machiavelli to examine the models for governing science presupposed by internalist and externalist historiographies, he would interpret both the centrality of rational persuasion in the former and the focus on coercion in the latter as fundamentally antipolitical. Here politics is more than just the effective mobilization of people toward some end; in addition, the people in question must benefit enough from the process that they are willingly mobilized in the future. As Machiavelli saw it, the Prince who failed to attend to his subjects' needs failed as a politician and thus no longer deserved to rule. Let us keep this sense of the political in mind for what follows.

Rational persuasion, by contrast, appears to inhibit any efforts at mobilization by presuming something that history has repeatedly shown to be false: namely, that there is a way of presenting the facts of a case so that the people will freely choose the most rational course of action. Machiavellians vary in their diagnosis of the failure of persuasion: Some trace it to fundamental human irrationality, others to the unfeasibility of a rational forum in a large heterogeneous society (Fuller 1988a). In either case, the result is the same: endless, debilitating wrangling, which indirectly promotes coercion, the other form of antipolitics. If persuasion purports to change the intellect without imposing on the will, then coercion operates in reverse by forcing a change of will while leaving the intellect largely unaffected; hence, the coerced feel oppressed by and fearful of the coercers. Coercion undermines politics by promoting self-consuming forms of power, whereby the force used to mobilize people on one occasion is enough to ensure that they cannot be mobilized again without applying still more force. As opposed to all this, manipulation is consummately political. Both the will and the intellect are changed, so that the manipulated internalize the interests of the manipulators as a natural extension of their own interests (Strauss 1958). When scientists impose negative sanctions on each other, they are engaging in mutual manipulation of this sort.

Critics of the sanction interpretation of scientific norms are right in claiming that the scientific community is not a particularly coercive one. After all, the aims of science cannot be promoted unless scientists are receptive to the idea of changing their minds; yet it is the unfeasibility of effecting such a change that has perennially led politicians down the path of coercion. But to reveal the inadequacy of coercion as a means of governing science is not necessarily to vindicate rational persuasion. Over the course of history, knowledge has been pursued by an increasing number of ever more disparate people, each of whom has realized that there is no a priori guarantee that any conclusion she draws will be noticed by her fellow inquirers. If the pursuit of knowledge were merely a matter of contributing to a databank, then inquirers might be satisfied with making their contributions anonymously, secure in the understanding that if anyone were interested in a conclusion they had reached, it could be retrieved from the databank. However, most of what an inquirer does when pursuing knowledge is directed toward another goal, namely, getting her fellows to find her work central to their concerns. In other words, the inquirer must use her research as a vehicle for forging a common identity with a target group of colleagues; otherwise she is open to the true sense in which negative sanctions operate in science, namely, in the process of being ignored or, worse yet, not being talked about.

Political theorists have often remarked that the difference between an authoritarian and a democratic regime lies in the fears that each sort of regime instills in the populace. Whereas citizens of an authoritarian nation fear pain inflicted by the sovereign, those of a democratic rule fear isolation from their fellows (Noelle-Neumann 1984, ch. 3). Philosophers, and even some sociologists, of science persist in conceiving of normative constraint in purely authoritarian terms, which leads them to imagine norms as being regularly enforced from some more or less well-defined power source external to any individual scientist's will (e.g., the sovereign gatekeepers). Scientific innovation is then typically portrayed in terms of the innovator heroically refusing to accept these constraints, which often culminates in a head-on confrontation between the individual and the establishment, as in the case of Galileo vs. the Church. Paul Feyerabend (1975) has perhaps been most diligent in siding with the lone innovator as operating in this authoritarian picture, the surreal consequences of which should be enough to empirically discredit the entire picture.

But none of this is to downplay the presence of constraints in science. On the contrary, my point is only that these constraints are largely self-imposed by individuals who realize that their social survival depends on the recognition they can gain from others. As research into public opinion and mass communications has shown, this is the sense of social control that operates in democratic regimes, whereby "power" is a matter of second-guessing peers rather than obeying superiors, and "constraint" is more an enabling than an inhibiting condition of action (cf. Park & Burgess 1921, chs. 4, 7, 12). Innovation then becomes an attention-grabbing process in which all scientists are regularly engaged. However, each scientist imposes limits on her own innovativeness as a function of her ability to gauge what is (and is not) likely to catch on within the time frame in which her continued survival in the field will be determined. In this respect, scientific innovation should be susceptible to the sorts of entrepreneurial and market analyses that have already been fruitfully used to explain technological innovation (Elster 1983).

12. If the Display of Norms Is So Disparate, Then the Search for Cognitive Coherence Is Just So Much Voodoo

In discussing the divergent modalities for expressing community norms, I argued that the way in which a scientist comes to assert that, say, the falsification principle governs her professional activity (via the study of methodology) is quite distinct from the way in which the scientist comes to behave in a manner that an observer would regard as falsificationist (via her apprenticeship in the lab). Moreover, if the knowledge stored from these learning experiences is as encapsulated as the experiences themselves are distinct, then there is little reason to think that when the scientist is asked to state the principles governing her professional activity, she is doing anything more than simply accessing what she has been trained to say when asked such questions. In particular, there is little reason to think that, prior to answering, the scientist inductively checks her verbal knowledge of the falsification principle against the record of her own experience.

If philosophers were not typically saddled with a view of the mind's mechanics that aspires to (and often seems to achieve) integration, then the unfeasibility of cross-validating our memories of words and deeds would become more apparent. After all, the longest pedigree (over one hundred years) in the history of experimental psychology—from Hermann Ebbinghaus on the rate of forgetting to Tversky and Kahneman on cognitive heuristics—belongs to demonstrations of the natural *disanalogies* between human memory and any reliable record-keeping device (cf. Rosenfeld 1987). Be that as it may, the even more venerable method of *reflective equilibrium* obscures this tradition of psychological findings by making it seem as though there is some principled means of integrating the data in our various memory stores (i.e., intuitions, articulable beliefs, dispositions to act, etc.) *and* that somehow our epistemic judgments are improved in the course of attempting such an integration (Daniels 1980; Thagard 1988, ch. 7). It is curious that both the desirability and the feasibility of reflective equilibrium have been subject to little scrutiny (a notable exception is Stich 1988). I will start by challenging the desirability of reflective equilibrium by undermining the connection it presupposes between improving scientists' minds (i.e., making them more self-consistent) and improving the science that is done.

In Chapter Three, I briefly considered the Faust (1984) and Stich (1985) strategies for a sociology of rationality. The difference between these two strategies is at the heart of an essential ambiguity in the attempt to reconcile the normative and naturalistic dimensions of epistemology. Is a philosophical theory of rationality supposed to *improve* the production of knowledge, or rather, is it supposed to *explain* why knowledge production works as well as it does? There are several ways of characterizing the distinction drawn here. If we conceive of knowledge production as a system, then one obvious way is in terms of Dennett's (1978, ch. 1) division of stances: intentional vs. design. For example, adopting the intentional stance, Faust interprets the knowledge system as trying to produce reliable representations of reality, but he notices that it is only partially successful at this task, largely due to the suboptimal social arrangement of the system's component parts, namely, cognitively limited knowledge producers. Clearly, then, Faust starts by attributing a goal to the knowledge system and then suggests ways in which the parts may improve their performance toward realizing that goal.

In contrast, Stich is already satisfied with the quality of products that the knowledge system normally generates. For him, the design stance is more appropriate, as he wants to account for how the parts contribute to the system working as well as it does. Whereas Faust's explicit advice on corrective procedures gives his theory of rationality a more "normative" look, Stich appears more the "naturalist" because he already finds rationality implicit in the ongoing activities of the knowledge producers. In the preceding section, I raised a mixed strategy, manipulation, which works, not by issuing methodological directives to alter the knowledge producers directly, but by indirectly altering the environments in which the knowledge producers labor, until the desired forms of knowledge are produced.

Nevertheless, as long as manipulation is not taken seriously as a philosophical issue, the main source of philosophical ambivalence remains. Should the naturalized epistemologist be doing social policy (cf. Faust) or social science (cf. Stich)? Although most epistemologists would claim to want to improve knowledge *products*, they normally attack the issue by proposing theories of rationality that are designed to improve the knowledge *producers*. One striking symptom of this strategy is the perennial demand that the rational knowledge producer be, if nothing else, self-consistent. And when treated as a technical project, the goal of self-consistency is the achievement of reflective equilibrium, whereby the individual does what she thinks is right, believes what she thinks is true, and mutually adjusts her ethical and epistemic norms to form a coherent world-view. Given the call for self-consistency in all the branches of normative philosophy, it is significant that no *empirical* case has yet been mounted for thinking that a self-consistent cognizer, or one striving for self-consistency, is more likely to advance the frontiers of knowledge than a cognizer oblivious to the demands of consistency (as, I believe, we normally are, until called upon to account for our activities), or one who admits inconsistency but fails to do anything about it. At most, what might follow about a cognizer who ignored consistency constraints is that she would never know by exactly how much she has advanced the frontiers of knowledge. Not surprisingly, the demands of self-consistency seem most pressing when the cognizer is treated as what Carnap and Fodor have called a "methodological solipsist" (cf. Fodor 1981).

From this individualistic standpoint, cognitive norms are pre-scribed without an examination of the likely epistemic consequences of several agents' concurrently conforming to those norms. However, the severance of self-consistency from its epistemic consequences has probably had the profoundest effect in the concept of criticism presupposed in analytic philosophy today, which is quite unlike that found in the sciences. Considering this concept in some detail will give us a glimpse as to why philosophy never seems to make any genuine progress.

If the scientist is a bounded rational agent, as was suggested in Chapter Two, then one would expect that even criticism, frequently regarded as the key to the growth of knowledge, ought not to be pursued as an end in itself. Indeed, the epistemic effectiveness of criticism is inherently self-limiting. In other words, there is reason to think that criticism (like other key epistemic practices) is most effective in producing knowledge when it is deployed judiciously, not when it is deployed to its fullest extent. By contrast, philosophers who adhere to the internalist myth tend to believe that if scientists did not have to worry about the relevance of their immediate concerns to some larger extra-scientific context, they would simply be trying to find the best possible solution to a problem that interests them. Although that might well be true, it is not clear that the net result of these impossibly idealized conditions would be anything like what we normally take to be epistemically progressive about scientific pursuits. Indeed, under such conditions, science might develop a mean skeptical streak and start to look like medieval scholasticism or (dare I say) contemporary analytic philosophy. Let us now consider what all this means.

Popper (1963) has been the most eloquent advocate for a sentiment widely held among philosophers about the critical attitude as the link between the methods of philosophy and science. In brief, science, especially in its more revolutionary phases, is held to be the continuation of Socratic questioning by more technical means. In the twentieth century, the logical positivists and their analytic-philosophi-cal offspring have deliberately tried to bring their practice into line with this sentiment. The result has been that philosophy—at least as practiced in most of the English-speaking world—is the only discipline in which adeptness at criticism is considered the premier professional virtue (Rorty 1982, ch. 12). Whereas experimental elegance and generalizability are leading virtues in the sciences, erudition and insightfulness win plaudits in the humanities.

This is not to say that these other disciplines do not value a critical attitude in their practitioners. In fact, studies show that such an attitude is one of the main criteria that scientists cite for judging their colleagues as intelligent (cf. Mitroff 1974). However, maximizing the display of one's *intelligence* is one thing; maximizing the growth of *knowledge* is something else entirely (cf. Elgin 1988). And as for the intuition that indefinitely enhancing each inquirer's critical sensibility will proportionally increase the quality or quantity of the knowledge that the inquirers collectively produce, that idea sounds too much like the fallacy of composition to be trusted. Thus, there may well be a level beyond which criticism starts to become counterproductive to its stipulated end of promoting the growth of knowledge. Yet, as we will now explore in more detail, it would seem that Anglo-American philosophy since the positivists and Popper has fallen precisely for this fallacious inference.

No doubt, because the philosopher's most primitive reflex is to criticize, it has taken two social psychologists, William Shadish and Robert Neimeyer, to remind us that criticism can be institutionalized on a variety of levels (Neimeyer & Shadish 1987, Shadish 1989). Once positivism shifted the *topos* of philosophical argument from self-contained "systems" to common "problems" (Hacking 1984), criticism became subtly caught in a paradox of levels, which has since foreclosed any chance of real progress in analytic philosophy. On the one hand, a palpable sign of progress in the natural sciences is that, over time, the relevant units of criticism, such as particular claims and arguments, become smaller and more focused. This is the phenomenon that Kuhn dubbed "normal science," which presupposes that a community of inquirers share enough assumptions that they can devote their energies to solving well-defined puzzles. On the other hand, analytic philosophers have generally followed Carnap's (1956) lead in holding that criticism can be rationally applied to issues raised *within* a conceptual framework, relative to its own standards, but not to issues raised *about* a framework, relative to some "transcendent" or "metaphysical" standards. Now combine these two points with an awareness that major framework questions remain unresolved in philosophy, and the result is the pattern that criticism takes in contemporary analytic philosophy. It is a pattern quite unlike what one finds in the natural and social sciences. Analytic philosophers diligently solve puzzles even if they remain unconvinced that these puzzles are situated within the epistemically soundest framework available. Indeed, both Carnap and Popper are, strictly speaking, irrationalists when it comes to evaluating alternative conceptual frameworks prior to one's being adopted. This is the task that Nickles (1987) has called "heuristic appraisal," which figured prominently in my definition of the scientist as a bounded rationalist in Chapter Two.

Critical pieces are published more readily in analytic philosophy than in the sciences, but only if they are of a certain sort. Since framework issues are presumed to be unresolvable, articles must be criticized in the author's own terms, however foreign they may be to the critic's native idiom. Thus, it is routine for, say, an a priorist epistemologist to criticize one of a more naturalistic bent without invoking specifically a priori considerations, instead restricting herself to problems with the "internal coherence" of the naturalist's position. Scientists will look in vain in these pieces for what they would recognize as "external validity checks": such as, mention of non-philosophical facts to which philosophers can safely hold each other accountable, a sense that the history of the debate has given one side or the other a greater burden of proof to bear, or even an assessment of the compatibility of a position with developments in other branches of philosophy (Nagel 1986 is an important exception here). Indeed, as Dennett (1987a, ch. 8) has astutely noted, even the analytic practice of appealing to intuitions about hypothetical cases should not be seen as drawing on some deep-seated beliefs that philosophers can be expected to share, but rather on the much less exalted ability to draw similar conclusions from similarly described cases. The critic appeals to intuition, not so much because she holds the intuition itself to be of general significance, but because she holds it to provide a counterargument to what her opponent wants to maintain (cf. Cohen 1986, part 2).

The difficulties that the scientist faces in understanding the analytic philosopher's critical sensibility highlights the extent to which philosophy's disciplinary purity is very much an artifact of enforcing certain writing conventions, which are, in turn, born of certain notions about the limits of criticism—namely, to matters internal to a position or an author's perspective. It is a testimony to the continuing influence of Bertrand Russell's realist philosophy of logic that philosophers regard critiques of internal coherence as having a more "objective" or "universal" character than critiques based on the empirical status of claims entailed by a position (Passmore 1966, ch. 9). To get a little perspective on this matter, keep in mind that when formal logic was first introduced by George Boole as a critical calculus in the nineteenth century, it was generally reviled as just another weapon in the sophist's arsenal, a weapon that would allow the philosopher to make or break arguments without having to hold any substantive views of her own, since any check for internal coherence would be specific to the position under study (Passmore 1966, ch. 7). Thus, a brilliant critique of one position would not even have to imply that the critic is unsympathetic to the criticized position, but merely to its particular formulation. Indeed, it was feared (by Mill, William Whewell, and other philosophers of science) that a promiscuous use of formal logic would engender a new era of medievalism in knowledge production, whereby generations of formal critiques would yield a stockpile of internally coherent, albeit byzantinely constructed, systems, incommensurable with one another and forever in play (Fuller 1985).

Why, for all their prowess in practice, do philosophers (except Bartley 1984) fail to have much of theoretical interest to say about the nature of criticism? One reason is that they are not sensitive to what may be called the *asymmetry of intention and reception*. Philosophers naively think that if a blunder has been made in a paper, and the critic pinpoints it as clearly as possible, then, assuming the rationality of the paper's author, the author will either admit to the blunder or try to show that the critic herself has made a blunder. This view presupposes that one ought to be as exact and thorough in one's criticisms as possible. If criticism were nothing more than the *identifying* error, then this view might be sound. However, as was noted in Chapter Three, the function of criticism in knowledge production also involves *eliminating* the error, to say nothing of replacing it with the truth. This means that the author of the paper must be well-disposed to receiving criticism, which, in turn, may require that the criticism be pitched somewhat indirectly.

We must admit that little research has been conducted on the relevant sense of "indirectness" involved here, but we may take a couple of hints from less-exalted studies on how students of English composition react to criticism of their papers. "Conscientious" instructors who detail every fault tend to meet with less success in student revisions than instructors who focus on one or two faults per paper (Hirsch 1977, ch. 7). Now apply the student's implicit psychology to an academic author faced with a journal referee's critique. If the critic isolates every fault in the paper, the author might well turn her off entirely or make only the most cosmetic adjustments for purposes of publication. However, if the critic aims at only a few general targets, then the author would probably take the criticism more sympathetically and, in the course of doing so, end up altering more of her argument than she otherwise would. In short, the maxim of epistemic criticism may be this: if you want to eliminate *any* errors, then you must be prepared to ignore *most* of them.

Having cast aspersions on the *desirability* of the unlimited search for self-consistency, the hallmark of reflective equilibrium, let us now to turn to another dimension in which it is equally dubious: *feasibility*.

Reflective equilibrium is a voodoo process because it treats the highly labile, encapsulated, and protocol-sensitive character of our knowledge bases as though they were stable chunks whose contents can be surveyed and compared before the mind's eye. On the contrary, when (and if) a scientist compares her knowledge of the falsification principle with her memory of falsifying experiences, it is probably as an implicit exercise in social control. For if the scientist concludes that she has not typically acted in accordance with the falsification principle, then it is not the verbal formulation of the principle that is adjusted to match her experience, but rather her experience is deemed deviant or somehow unrepresentative of the norm expressed by the principle. (At the very least, the scientist assumes the burden of proof to show otherwise.) It may turn out that an unintended consequence of many scientists routinely discounting their personal experiences in this fashion is that, to an observer, a wide but unsystematic disparity emerges between what the scientists say and what they do: They say similar things while doing dissimilar things. Once again, this would go to show that the scientist's judgment does not necessarily rest on reliable information about how scientists tend to behave, but simply on the way she was taught to compare methodological dicta with whatever experiences she may retrieve from memory.

It might be objected that I have shortchanged the integrative powers of global principles of rationality. True, a principle like falsification can be invoked to frame and assess the normative status of any particular epistemic experience we might have. But by no means does this fact imply that the principle summarizes our actual epistemic experience, however we may choose to record it—either through memory or through some more reliable means. However, it may be objected that even if true, the internally divided character of the mind is not enough to show that reflective equilibrium could not bring a measure of coherence to our disparate data stores. Indeed, reflective equilibrium may be especially well suited for such a model of the mind, since a mind in deep disunity is less likely to reach "equilibrium" prematurely, which, in turn, provides a greater opportunity for "reflection" to rationally refashion our mental machinery for future use. Both Fodor (1987) and Dennett (1984) go so far as to argue the panglossian thesis that such disunity provides exactly the right psychological opportunity for consciousness, or "concentration," as Fodor prefers to call it.

But would that our epistemic situation be as straightforward as our two cognitive scientists suggest! On the one hand, social psychologists who have studied the resolution of cognitive dissonance *do* perceive a mental tendency toward "consistency." On the other hand, consistency is typically achieved by a method that is less misleadingly described as "reflective *disequilibrium*" (cf. Feldman 1966). When people notice a discrepancy between (say) their avowed principles and their actions, they do not remedy the situation by adjusting their beliefs to conform to their actions (or vice versa). Instead, they typically conjure up some overlooked factor (perhaps mitigating circumstances) that renders the discrepancy more superficial than it first appeared, in other words, ignorable upon further reflection; hence, reflective disequilibrium. Of course, the overlooked factor would not have been invoked had the discrepancy not been noticed, but since invoking the factor results in regaining cognitive consistency with minimum effort, there is little incentive for probing into the ad hoc nature of this factor.

From the standpoint of the scientific observer, the agent appears to be decreasing the opportunities for the mutual cross-examination of her avowed principles, intuitions, actions, and evaluations by interpolating zones of casuistry, which are designed to turn most of the conflict that would emerge in such a cross-examination into exceptional, and therefore excusable, cases. From the standpoint of the agent, however, the process of reflective disequilibrium is hardly ever so self-conscious. (It is noteworthy that the Catholic Church has traditionally been vividly aware of the need for casuistry to retain the integrity of its avowed principles in a less than perfect world, cf. Jonsen & Toulmin 1988.) Through some selective forgetting, most of the disagreeable cases drop out of easy retrieval, while the rest are assimilated to normative stereotypes, which together serve to render the agent's self-image coherent.

In considering why people cannot be easily convinced that they do not live up to their own moral principles, the social learning theorist Albert Bandura (1969) has observed that the agent's casuistic powers are never far from the surface of consciousness. Bandura's point, which I claim equally applies to scientists vis-a-vis their professional norms, is summarized in the slogan "self-disengagement breeds self-justification." In short, people can maintain good self-images more easily if they do not monitor their own activities too closely and if they then show, when questioned, that this proved to have been a good strategy since the apparent discrepancies between their words and deeds can be explained away as exceptional cases.

But still, the defender of reflective equilibrium may persist: Is not enough already known about how reflective equilibrium *ought* to work, as well as about how our minds naturally fail to work, that people could be forewarned about their liabilities while being trained to maximize the right sort of cognitive consistency? In Bandura's terms, the scientists should learn to be more discriminating self-regulators. But even assuming that the evidence does support the epistemic utility of the new and improved form of reflective equilibrium, I would then question the extent to which its strategy of attuning the cognizer to discrepancies by adopting an external standpoint would be anything more than an elaborate exercise in biofeedback and behavior modification, areas where philosophical introspection has been all but eclipsed by empirical science. The point, then, is that were we truly serious about achieving the *goals* of reflective equilibrium, we would do best to turn away from the usual Cartesian methods and toward more experimental ones (cf. Churchland 1979; Lyons 1986).

Needless to say, most philosophers—even naturalistic ones—are unlikely to regard biofeedback and behavior modification as legitimate successors to reflective equilibrium. But readers who are inclined to accept the heirs apparent may appreciate the irony involved in the fate that would now befall the philosopher's preferred persona, the skeptic, who, in all her doubts, never countenanced the possibility that an evil demon in the social sciences would someday reveal that the problem of knowledge of the external world is itself predicated on a false theory of the mind! Hilary Kornblith (1988) has noted that the skeptic's problem cannot get off the ground unless it is granted that we have determinate beliefs and that we know which beliefs we have—an epistemic state that is then made to stand in sharp contrast to our ignorance of the truth or falsity of those beliefs. However, the skeptic's assumption is an empirical one about the degree of epistemic access that we have to our own mind vis-a-vis to what lies beyond it, an assumption that on the weight of current psychological evidence is probably false!

In conclusion, let us recall the various strategies available to social scientists for explaining the discrepancy between a subject's avowed principles and actual behavior. The social scientist may claim:

(i) that the subject does not behave according to her own principles;

(ii) that the subject has a false understanding of the principles governing her behavior;

(iii) that the subject's avowal of principles should be treated, not as a putative account of her own behavior, but simply as more first-order behavior to be scientifically explained.

Missing among these strategies is any recourse to the philosopher's strategy of reflective equilibrium, which would have the social scientist presume that avowals and actions are equally valid expressions of the agent's underlying beliefs and desires. The philosopher would then aim for an account in which most of these beliefs turn out to be justified. This may require showing that the agent's principles are somewhat crudely stated or that some of her behavior was indeed wayward, but none of this would deny that a coherent story could be told of the interplay between the two normative modalities in expressing a common set of beliefs. I submit that reconciling the modalities in this way would strike the social scientist as reminiscent of the eighteenth century chemist who thought that there was a subtle but coherent story to be told about how phlogiston could have positive weight when released from wood, yet negative weight when released from metallic ores. As even most phlogiston chemists had come to realize during this period, either phlogiston is present in only one of these cases and not the other, or the entire concept is sufficiently confused to call for a radical re-analysis. Clearly, history plumped for the latter option. I am betting that it will do so again.

Epistemic Autonomy as Institutionalized Self-Deception

Although autonomy is nowadays taken to be the cornerstone of any principled action, either in morality or methodology, this is largely an eighteenth century innovation, grounded in the possibility of a self-legislated microcosm of the greater universe governed by the Creator (who, by this time, was armed with Newton's Laws; cf. Schneewind 1987). In order to preserve free will in a deterministic world, it was important to carve out a realm, however small, whereby the agent could countermand dominant natural tendencies with laws of her own creation. Thus, the need for autonomy arises in a world where people feel that they are already under a good deal of metaphysical or social control, which leads them to look for an area of free play.

This point is vividly illustrated in autonomy's original context, as the Greek *autonomia*. The Stoic Epictetus, a model for Nietzsche's "slave morality," is the key witness here. Epictetus, himself a Roman slave, argued that, despite his physical oppression, his oppressors still could not control what he thought unless he allowed them. That sphere, in which Epictetus' will remains decisive, is the self-legislating soul. Now, if it turns out that the Romans can get Epictetus to do whatever *they* want, since they are only interested in him as a vehicle of hard labor, then Epictetus' claim to autonomy will look like a rationalized retreat, a special case of preferences (i.e., the realm that Epictetus would like to control) being adapted to match expectations (i.e., the realm that Epictetus could probably control). The point, then, is that people can easily be made to feel autonomous if the sphere of their lives that they wish to control is limited enough (cf. Foucault 1977). When this point is applied to the history of knowledge production, a variety of suspicions are raised: Who is promoting an ideology of autonomy at the time? Exactly over which aspects of the knowledge producers' activities? And who stands to gain by an acceptance of the ideology?

To suspect a claim to autonomy as being a disguised instance of what Jon Elster (1984b) calls an "adaptive preference formation" is to suggest some major rewritings of the history of science. For example, as was noted in Chapter Two, at its founding, the Royal Society of Great Britain, the first autonomous scientific body, was granted a charter on the condition that it did not inquire into what we would now class as the human sciences: religion, metaphysics, rhetoric, politics. It would be interesting to see if, and when, a collective "sour grapes" set in, whereby members of the Society started arguing that it was just as well that they did not conduct experiments in human affairs, given their intractably complex character. If so, their sense of autonomy would increase, as they would no longer have an interest in doing something that had been rendered impracticable for them. In terms of the thesis advanced in this book, I would guess that our intuitive sorting of "internal" and "external" influences on science reflects another type of adaptive preference formation. The philosopher (or the scientist, for that matter) who does not feel that the scientist's autonomy is restricted by her not having control over the sources of research funds probably does not consider funding an inherent feature of the scientific process, and would therefore not include it in an internal history of science. By contrast, a Marxist analysis would reveal the very same scientist to be much less autonomous, since economic factors would be defined as integral to science, and hence as something over which the scientist ought to have control but in fact does not. The issue here becomes murky in the long run, as scientists start to unwittingly align their cognitive interests with their funding potential. Thus, successful scientists come to provide "epistemic" reasons for preferring to do the sort of research that just so happens to be the sort most likely funded (a phenomenon that Elster tags "sweet lemons"), whereas unsuccessful scientists chide the successful ones for allowing their research trajectories to be diverted by such "external" matters as money.

The relevance of these points becomes especially striking when we look at the sense of autonomy that was attached to the guild right of "freedom to inquire" (*Lehrfreiheit*) in the German academic community at the end of the nineteenth century, the model of academic freedom in our own times (Hofstadter & Metzger 1955; Proctor 1991, ch. 10). Thus, inquirers have been encouraged to follow their own leads, which has, in turn, given each inquirer greater freedom over an ever shrinking domain of knowledge. And, as I have tried to show in this book, this point applies no less to the philosophy of science. The problem, of course, is that as responsibility for charting the overall direction of research has been abdicated, the tasks associated with that responsibility have been passed over, almost by default, to university administrators, systems analysts, and state legislators. Without questioning the worthiness of these people to the task of managing the knowledge industry, the fact that we look with disdain upon the "intellectual" merits of their activities shows just how alienated our epistemic practices have become from the point of those practices.

Big Questions and Little Answers— A Response to Critics

Q: *Why do we need a normative discipline, especially one as heavy-handedly prescriptive as social epistemology? Isn't the whole idea behind "going naturalistic" to let empirical reality, rather than our preconceived notions, dictate appropriate courses of action?*

A: All disciplines are normative, though rarely self-consciously so. A discipline often doesn't seem normative because its members appear to be satisfied with the course that their activities are collectively taking (or at least there are no centrally located forums for voicing dissatisfaction). By contrast, a self-consciously normative discipline is one that makes the ends of its inquiry an ongoing subject for negotiation. In that sense, any such discipline already practices social epistemology. The naturalistic turn helps by bringing to light discrepancies between what is and what ought to be the case. The naturalist then has the option of either bringing the ideal closer to the real or bringing the real closer to the ideal (cf. Canguilhem 1978). I prefer the latter, though as I observed in the Coda, scientists have often found the former option the path of least resistance. In either case, naturalism forces us to see that *work* is required to make up the difference between "the ought" and "the is."

Q: *Since so much of this book is devoted to the major meta-problem of naturalism—namely, how to integrate research utilizing divergent empirical methodologies—hardly anything is said about the sorts of substantive empirical findings that would inform the ongoing discussions about the "ends of inquiry." What should we be looking for?*

A: This is an eminently fair question, since I don't believe that "knowledge for its own sake" is anything more than a thinly disguised plea for inquirers to have complete discretion over the course of their inquiries. As there are no natural ends to inquiry, ends must be judged in terms of consequences. However, as every ethicist knows, the consequences of a given action are incredibly difficult to trace, even supposing that we have an adequate theory of causation: How does one circumscribe the range of relevant past causes and future effects, especially for something as ontologically elusive as knowledge? One quick and dirty way favored by many sociologists is to count as knowledge whatever passes as knowledge in a particular community. The rationality of the community can then be assessed according to whether the knowledge produced by its members enables them to get what they want. Moreover, on this view, we don't need to take a stand on the kind of "thing" knowledge is because we simply take the community's word for it. Although I have been attracted to such a strategy in the past, the economic and legal issues surrounding intellectual property, that ultimate embodiment of knowledge, has made me increasingly suspicious of the sociological formulation (cf. Fuller 1991b, 1992c).

My suspicions can be summarized in the following question: Are the members of a particular discipline the only people who should be involved in negotiating the ends of that discipline's inquiry? True, they are the ones who determine what passes as knowledge in their community, but that fact presupposes a politically and economically supportive environment (typically the unquestioned redistribution of income from public taxation) combined with the inquirers' discretionary control over how the fruits of their labors are represented to the public. These background conditions constitute what economists call "externalities'—that is, the hidden costs and benefits that are required for the discipline's normal operation. Social epistemology holds that a theory of knowledge is not adequate unless it identifies and accounts for these externalities. It is difficult to see how this can be done without relinquishing the standpoint of any given community of inquirers. My model here is intellectual property law, which is designed to "internalize" externalities by incorporating the interests of third parties who might be affected by the introduction of a certain form of knowledge. For example, patents last only a fixed number of years so as to reward one person's inventiveness now without discouraging others from inventing in the future. My objective here is to articulate principles of what may be called *epistemic justice*, namely, a fair representation of the different stakes that people have in the knowledge production process. It is worth remembering that, as knowledge production becomes more specialized, we are all third parties to an ever larger share of the process.

Talking about "the ends of inquiry" in this way has made several classical epistemologists wonder whether I am really talking about knowledge in strictu sensu, namely, as justified true belief, a state of mind rightly aligned with external reality. Likewise, if in a somewhat more empirical vein, a locution like "matching an hypothesis up against an observation" contributes to the idea that the ends of inquiry is akin to a target one is trying to hit. While philosophers like to identify such a target as the "transcendental limit" on all inquiry, empirically speaking the target imagery is probably better suited to more mundane epistemic practices, such as formal examinations and other rites of professional passage. As I pointed out in Chapter Three, the target is typically an artifact, such as the computer, which becomes the standard against which human performance is judged and shaped. When this simple fact is forgotten, and the computer is taken to be an "instantiation" of some transcendent qualities, we enter the realm of what the postmodernist philosopher Jean Baudrillard calls "the hyperreal," whereby the simulation is taken to be more real than the empirically real things surrounding it. Classical epistemology and much of cognitive science typically dwell in hyperreality. I do not necessarily mean this as a criticism, simply an observation.

Another way of empirically exploring the ends of inquiry is to examine the impact of certain modes of inquiry on the lives of the inquirers, along the lines suggested by "virtue theory" (cf. MacIntyre 1984): Does humanistic or scientific training make one better able to flourish among one's fellows, including those without such training? Pragmatism's construal of ends promotes a normative constructivist study into the extent to which certain forms of knowledge enable people to get what they want. However, from the social epistemologist's standpoint, the deepest form of knowledge is that which enables a transformation of the knowledge production process itself. Thus, one is no mere bearer or consumer of knowledge, but a full-fledged producer capable of dictating the ends of the process. One field of study currently underutilized by epistemologists which could aid in fleshing out these research agendas is *program evaluation*. A founding father of this field is the philosopher of science Michael Scriven, who first advanced the idea suggested here, that the express goals of an organization, such as a discipline, may not be the best way to evaluate its performance (cf. Shadish et al. 1991, esp. ch. 3).

Q: *Doesn't social epistemology suffer from the same authoritarian—if not totalitarian—impulse of philosophy, namely, wanting to dictate norms from "on high" without considering the interests of local knowledge practitioners?*

A: Social epistemology sounds most normative when it speaks the language of legislation, as in the idea of deriving "principles of epistemic justice." Unfortunately, talk of legislation still carries the unsavory seventeenth century connotation of an absolute sovereign legislator. Even modern positivist theories of law typically postulate the fiction of a supreme legislator, whose absolute ability to issue sanctions ensures the force of law. Critics wonder: Can Plato's philosopher-king be far behind? In response, I would draw a distinction between the philosopher's responsibility to *introduce* normative considerations and her audience's responsibility to *resolve* them through some appropriate "legislation," broadly construed. My latest book tackles this thorny problem, and the theory of rhetoric that it requires (Fuller 1992a). The social epistemologist cannot get the normative side of her project off the ground unless she cultivates *ethos* and *kairos* with her audience. In other words, to get their attention, she must show that she understands their interests, and to get their action, she must show that they have a stake in the fate of her proposal.

Some critics suggest that I replace my "top-down management" style with something more "ecologically" or "locally" sensitive to the actual needs of particular knowledge communities. Given what we know about the disparateness of disciplinary practices (much of which I had previously brought to philosophical attention), why do I now want to sound like a positivist legislator? First, it is worth noting that the epistemic differences we detect among communities presuppose a fairly large-scale, perhaps even global, perspective from which systematic comparisons can be made. If we were as locale-bound as many postmodernists claim, then we would never have learned that there were distinct locales in the first place! The issue, then, is what do we make of these differences. Are they to be left alone, as if they were natural, or are they to be treated in some other fashion? While I disavow the *universalist* dogma that all knowledge communities ought to behave in the same way, I am nonetheless a *globalist*, that is, I believe that as long as all of these communities inhabit the same planet, none of us can flourish unless we coordinate and, where necessary, resolve our differences. The relevant source of "pollution" here is not toxic wastes but proliferating disciplinary jargons. And here I am simply offering a politically more correct way of asserting the philosopher's professional obligation to unify people by challenging the ideas that divide them. Sensitivity to local differences is a necessary starting point, but *only* a starting point.

Q: *Social epistemology espouses an odd sort of naturalism, since it would reduce the natural sciences to the social sciences, and even then, certain social sciences, such as psychology, undergo substantial reinterpretation before being accepted into the fold. How does this square with the "naturalism" of other philosophers and sociologists?*

A: An interesting feature of the term "naturalism" is that while it is used by analytic philosophers to mark a brand of progressive theorizing that is open to empirical considerations, it is a term of contempt for Marxists, deconstructivists, and hermeneuticians, all of whom depict "naturalists" as naive and dogmatic science worshippers. To be sure, the continental critics of naturalism are uncharitable, but not unjustified. There has been a tendency for naturalists, especially when they fashion themselves "Philosophers of X" (where "X" is a natural science), to become immersed in the details of X, leaving their philosophical scruples behind. It is important, therefore, to take naturalism to be a commitment to a certain "scientific attitude" (which I tend to model on Popper's method of conjectures and refutations) rather than to the substance of particular sciences. Thus, a naturalized epistemologist should be prepared to argue that existing disciplinary boundaries do more to impede than to improve inquiry, and that the claims made by one discipline can be best explained by principles borrowed from another discipline. After all, the most lasting contribution of the sciences is not this or that finding but the continual overcoming of institutional inertia, even within science itself. And so, while philosophers may not dictate to the special sciences, they need not be underlaborers either.

Keep in mind that most of the philosophers and sociologists at loggerheads with one another were originally trained in the natural sciences, and, despite their radically different accounts of science, almost all of them claim to be pro-science and to have studied science "scientifically." This strongly suggests that naturalism is a many-splendored thing. To appreciate this phenomenon, replace "science" with "God" in the history of Christianity. The more positivistic philosophers of science are like the Roman Catholic priesthood (remember Auguste Comte's Positive Religion!) in their belief that methodological mediation is needed between scientists and the Science–God. The more constructivist sociologists of science are like the Protestants, especially Lutherans, who argue that no such mediation is needed, as the scientists directly experience the Science–God in their everyday practices. More recently, Bruno Latour and Donna Haraway have argued for the interpenetration of science and everything else on the basis of the empirical blurriness of science's boundaries. This may be seen as an analogue of Pantheism. My own view of the Science–God is closer to eighteenth century

Deism. (The reader can draw the relevant parallels.) In any case, there is not a single Atheist in the house!

Q: *How do I square my evident attraction to the critical style of philosophizing championed by Marxists and Popperians with the cognitive limitations critique of criticism that I lodge at the end of Chapter Four? Doesn't this undermine the project of social epistemology altogether?*

A: The important point here is that I am critiquing a particular form of criticism that may be called *opportunistic*. Thus, analytic philosophers are opportunistic because their criticism is typically parasitic on the assumptions of the person being criticized, assumptions to which the critic may have no particular attachment; hence, the historical connection I see with classical sophistry. By contrast, I endorse a *nonopportunistic* form of criticism (cf. Fuller 1992a, ch. 3). The nonopportunist doesn't criticize until she has found a position of her own. She then launches only criticisms that are compatible with her own working assumptions. I believe that the nonopportunist is the better critic because, whereas the opportunist simply dips in and out of other people's discourses, the nonopportunist cannot criticize without recasting her interlocutor's position in a more comprehensive light, namely, by revealing the conditions that enable them to have a genuine *disagreement*, as opposed to a mere diagnostic exercise. Others may wish to relabel opportunistic–nonopportunistic as, respectively, *neutral–interested*, if they don't see things quite the way I do.

Now there are also degenerate versions of opportunistic and nonopportunistic criticism. Analytic philosophy turns into degenerate opportunism once the critic believes that it doesn't make sense to articulate a position of her own because rival positions are fundamentally incommensurable. Thus, the philosopher stipulates that there is no chance of expanding the critical community so as to enable opposing standpoints to resolve their differences. For example, Laudan (1977) concedes to Kuhn the incommensurability thesis, but then says that research traditions can be compared by how well they solve the problems they set for themselves by their own standards. Laudan is clearly imagining a set of hermetically sealed traditions, which is not true to how traditions intertwine over time (cf. MacIntyre 1984), but it *is* true to how analytic philosophers treat one another's projects! The degenerate version of the view I hold is a critique that merely articulates its alternative perspective without ever directly engaging the opposition. One might say that this would be an orthogonal redescription (or represcription) of the reality that the interlocutor's position describes (or prescribes). Consider here the Frankfurt School's "critical theory" vis-à-vis positivism (cf. Adorno 1976). Such theorizing is degenerate in that it sacrifices the ultimate aim of nonopportunism by dwelling too much on how to be nonopportunistic. After all, nonopportunism presumes that the interlocutor *can* be drawn into one's own discursive universe. To dwell on how this might happen in the abstract—when one could actually be courting the interlocutor—is to cultivate a morbidly involuted form of reflectiveness. Now, incorporating the interlocutor can be quite tricky, as it involves representing what the interlocutor is talking about in one's own terms and then convincing the interlocutor that these new terms are better. Recall the earlier remarks on *ethos* and *kairos*. As difficult a rhetorical feat as it is, it is necessary for nonopportunism to succeed (cf. Fuller 1992a, ch. 3, for more details).

Bibliography

Ackermann, Robert. 1985: *Data, Instruments, and Theory*, Princeton: Princeton University Press.

Ackermann, Robert. 1988: "Experiment as the Motor of Scientific Progress," *Social Epistemology* 2:327–335.

Adorno, Theodor, ed. 1976. *The Positivist Dispute in German Sociology*, London: Heineman.

Agassi, Joseph. 1985. *Technology*, Dordrecht: Reidel.

Aitchison, Jean. 1981: *Language Change: Progress or Decay?* New York: Universe Books.

Alexander, Jeffrey, Bernhard Giesen, Richard Muench and Neil Smelser (eds.) 1986: *The Macro-Micro Link*, Berkeley: University of California Press.

Apel, Karl-Otto. 1985: *Understanding and Explanation*, Cambridge MA: MIT Press.

Arkes, Hal, and Kenneth Hammond (eds.). 1986: *Judgment and Decision Making*, Cambridge UK: Cambridge University Press.

Arnauld, Antoine. 1964 (1662): *The Art of Thinking*, Indianapolis: Bobbs-Merrill.

Aronowitz, Stanley. 1988: *Science as Power*, Minneapolis: University of Minnesota Press.

Ayer, A.J. (ed.). 1959: *Logical Positivism*, New York: Free Press.

Ayer, A.J. 1971 (1936): *Language, Truth, and Logic*, Harmondsworth: Penguin.

Bachelard, Gaston. 1984 (1934): *The New Scientific Spirit*, Boston: Beacon Press.

Baier, Annette. 1985: *Postures of the Mind*, Minneapolis: University of Minnesota Press.

Baigrie, Brian. 1988: "Philosophy of Science as Normative Sociology," *Metaphilosophy*, 19, 3/4.

Bandura, Albert. 1969: "Social Learning of Moral Judgments," *Journal of Personal and Social Psychology* 11:275–279.

Barnes, Barry. 1982: *T.S. Kuhn and Social Science*, Oxford: Blackwell.

Baron, Reuben. 1988: "An Ecological Framework for Establishing a Dual-Mode Theory of Social Knowing," in Bar-Tal & Kruglanski (1988).

Bar-Tal, Daniel, and Arie Kruglanski (eds.). 1988: *The Social Psychology of Knowledge*, Cambridge UK: Cambridge University Press.

Bartley, W.W. 1974: "Theory of Language and Philosophy of Science as Instruments of Educational Reform," in R. Cohen & M. Wartofsky (eds.), *Boston Studies in the Philosophy of Science XIV*, Dordrecht: Reidel, pp.307–337.

Bartley, W. W. 1984 (1962): *The Retreat to Commitment*, 2nd ed., La Salle: Open Court Press.

Bazerman, Charles. 1987: *Shaping Written Knowledge*, Madison: University of Wisconsin Press.

Bechtel, William. 1985: "Realism, Instrumentalism, and the Intentional Stance," *Cognitive Science* 9:473–497.

Bechtel, William (ed.) 1986: *Integrating Scientific Disciplines*, Dordrecht: Martinus Nijhoff.

Becker, Gary. 1964: *Human Capital*, New York: Columbia University Press.

Ben-David, Joseph. 1984 (1971): *The Scientist's Role in Society*, 2nd ed., Chicago: University of Chicago Press.

Beniger, James. 1987: *The Control Revolution*, Cambridge MA: Harvard University Press.

Berkowitz, Leonard, and Edward Donnerstein. 1982: "Why External Validity is More than Skin Deep," *American Psychologist* 37:245–257.

Bernal, J.D. 1969: *Science in History*, Cambridge MA: MIT Press.

Bernstein, Richard. 1983: *Beyond Objectivism and Relativism*, Philadelphia: University of Pennsylvania Press.

Bhaskar, Roy. 1979 (1975): *A Realist Theory of Science*, Sussex: Harvester.

Bhaskar, Roy. 1980: *The Possibility of Naturalism*, Sussex: Harvester.

Bhaskar, Roy. 1987: *Scientific Realism and Human Emancipation*, London: New Left Books.

Bijker, Wiebe, Thomas Hughes, and Trevor Pinch (eds.) 1987: *The Social Construction of Technological Systems*, Cambridge MA: MIT Press.

Biro, John, and J. Shahan (eds.). 1982: *Mind, Brain, and Functionalism*, Norman: University of Oklahoma Press.

Bloomfield, Brian (ed.) 1987: *The Question of Artificial Intelligence*, London: Croom Helm.

Bloor, David. 1976: *Knowledge and Social Imagery*, London: Routledge and Kegan Paul.

Bloor, David. 1982: "Durkheim and Mauss Revisited: Classification and the Sociology of Knowledge," *Studies in the History and Philosophy of Science* 13:267–298.

Bloor, David. 1983: *Wittgenstein: A Social Theory of Knowledge*, New York: Columbia University Press.

Boring, Edwin. 1957: *A History of Experimental Psychology*, New York: Appleton-Century Crofts.

Bourdieu, Pierre. 1975: "The Specificity of the Scientific Field and the Social Conditions of the Progress of Reason," *Social Science Information*.

Bourdieu, Pierre. 1981: "Men and Machines," in Knorr-Cetina & Cicourel (1981).

Boyd, Richard. 1979: "Metaphor and Theory Change," in Ortony (1979).

Boyd, Richard. 1984: "The Current Status of Scientific Realism," in Leplin (1984a).

Brandon, Robert, and Richard Burian. (eds.). 1984: *Genes, Organisms, and Populations*, Cambridge MA: MIT Press.

Brannigan, Augustine. 1981: *The Social Basis of Scientific Discoveries*, Cambridge UK: Cambridge University Press.

Brannigan, Augustine, and Richard Wanner. 1983: "Historical Distributions of Multiple Discoveries and Theories of Scientific Change," *Social Studies of Science* 13:417–435.

Brehmer, Berndt. 1986: "In One Word: Not from Experience," in Arkes & Hammond (1986).

Brown, Harold. 1977: *Perception, Theory, and Commitment*, Chicago: University of Chicago Press.

Brown, Harold. 1978: "On Being Rational," *American Philosophical Quarterly* 15:214–248. Brown, Harold. 1988a: "Normative Epistemology and Naturalized Epistemology," *Inquiry* 31:53–78.

Brown, Harold. 1988b: *Rationality*, Boston: Routledge & Kegan Paul.

Brown, Harold. 1989: "Towards a Cognitive of Psychology of What?" *Social Epistemology* 3:129–137.

Brown, James Robert (ed.) 1984: *The Rationality Debates: The Sociological Turn*, Dordrecht: Reidel.

Brown, Robert. 1986: *Social Laws*, Cambridge UK: Cambridge University Press.

Brunswik, Egon. 1952: *A Conceptual Framework for Psychology*, Chicago: University of Chicago Press.

Bryant, Christopher. 1985: *Positivism in Social Theory and Research*, New York: St. Martin's Press.

Buchdahl, Gerd. 1951: "Some Thoughts on Newton's Second Law of Motion in Classical Mechanics," *British Journal for the Philosophy of Science* 2:217–235.

Burge, Tyler. 1986: "Cartesian Error and the Objectivity of Perception," in Pettit & McDowell (1986).

Callon, Michel, John Law and Arie Rip (eds.). 1986: *Mapping the Dynamics of Science and Technology*, London: Macmillan.

Campbell, Donald. 1958: "Common Fate, Similarity, and Other Indices of the Status of Aggregates of Persons as Social Entities," *Behavioral Science* 3:1–14.

Campbell, Donald. 1969: "Ethnocentrism of Disciplines and the Fish-scale Model of Omniscience," in M. & C. Sherif (eds.), *Interdisciplinary Relationships in the Social Sciences*, Chicago: Aldine Press.

Campbell, Donald. 1979: "A Tribal Model of the Social System Vehicle Carrying Scientific Knowledge," *Knowledge* 2:181–201.

Campbell, Donald. 1986: "Science's Social System of Validity- Enhancing Collective Belief Change and the Problems of the Social Sciences," Fiske & Shweder (1986).

Campbell, Donald. 1987a: "Guidelines for Monitoring the Scientific Competence of Prevention Intervention Centers," *Knowledge* 8:389–430.

Campbell, Donald. 1987b: "Interview with Steve Fuller and the Social Epistemology Seminar," 22 November.

Campbell, Donald. 1987c: "Neurological Embodiments of Beliefs and the Gaps in the Fit of Phenomena to Noumena," in A. Shimony & D. Nails (eds.), *Naturalistic Epistemology*, Dordrecht: D. Reidel.

Campbell, Donald. 1988: *Methodology and Epistemology for Social Science*, Chicago: University of Chicago Press.

Campbell, Donald, and Julian Stanley. 1963: *Experimental and Quasi-Experimental Designs for Research*, Chicago: Rand McNally.

Canguilhem, Georges. 1978: *On the Normal and the Pathological*, Dordrecht: D. Reidel.

Capek, Milic. 1961: *The Philosophical Implications of Contemporary Physics*, New York: Van Nostrand.

Carnap, Rudolf. 1956: *Meaning and Necessity*, Chicago: University of Chicago Press.

Cartwright, Nancy. 1983: *How the Laws of Physics Lie*, Oxford: Oxford University Press.

Cassirer, Ernst. 1950: *The Problem of Knowledge*, New Haven: Yale University Press.

Churchland, Patricia. 1986: *Neurophilosophy*, Cambridge MA: MIT Press.

Churchland, Paul. 1979: *Scientific Realism and the Plasticity of Mind*, Cambridge UK: Cambridge University Press.

Churchland, Paul. 1984: *Matter and Consciousness*, Cambridge MA: MIT Press.

Churchland, Paul. 1989: *A Neurocomputational Approach to the Philosophy of Science*, Cambridge MA: MIT Press.

Churchland, Paul and Clifford Hooker (eds.). 1985: *Images of Science*, Chicago: University of Chicago Press.

Clifford, James, and George Marcus (eds.). 1986: *Writing Cultures*, Berkeley: University of California Press.

Cohen, L. Jonathan. 1986: *The Dialogue of Reason*, Oxford: Oxford University Press.

Collett, Peter. 1977: "On Rules of Conduct," in P. Collett (ed.), *Social Rules and Social Behavior*, Totowa: Rowman & Littlefield.

Collins, Harry. 1981: "Stages in the Empirical Program of Relativism," *Social Studies of Science* 11:3–10.

Collins, Harry. 1985: *Changing Order*, London: Sage.

Collins, Harry, 1990: *Artificial Experts*, Cambridge, MA: MIT Press.

Collins, Randall. 1975: *Conflict Sociology*, New York Academic Press.

Cozzens, Susan. 1985: "Comparing the Sciences: Citation Context Analysis of Papers from Neuropharmacology and the Sociology of Science," *Social Studies of Science* 15:127–153.

Culler, Jonathan. 1983: *On Deconstruction*, Ithaca: Cornell University Press.

Cummins, Robert. 1983: *The Nature of Psychological Explanation*, Cambridge MA: MIT Press.

D'Amico, Robert. 1989: *Historicism and Knowledge*, London: Routledge.

Daniels, Norman. 1980: "Reflective Equilibrium and Archimedean Points," *Canadian Journal of Philosophy* 10:83–103.

Danziger, Kurt. 1990: *Constructing the Subject*, Cambridge UK: Cambridge University Press.

Darden, Lindley, and Nancy Maull. 1977: "Interfield Theories," *Philosophy of Science* 44:43–64.

Davidson, Donald. 1983: *Inquiries into Truth and Interpretation*, Oxford: Oxford University Press.

Davidson, Donald. 1986: "A Nice Derangement of Epitaphs," in Ernest Le Pore (ed.), *Truth and Interpretation*, Oxford: Blackwell.

Dear, Peter. 1988: *Mersenne and the Learning of the Schools*, Ithaca: Cornell University Press.

De Mey, Marc. 1982: *The Cognitive Paradigm*, Dordrecht: D. Reidel.

Dennett, Daniel. 1978: *Brainstorms*, Cambridge MA: MIT Press.

Dennett, Daniel. 1982: "Making Sense of Ourselves," in Biro & Shahan (1982).

Dennett, Daniel. 1984: *Elbow Room*, Cambridge MA: MIT Press.

Dennett, Daniel. 1987a: *The Intentional Stance*, Cambridge MA: MIT Press.

Dennett, Daniel. 1987b: "Cognitive Wheels," in Pylyshyn (1987).

De Sousa, Ronald. 1987: *The Rationality of Emotions*, Cambridge MA: MIT Press.

Devine, Patricia, and Thomas Ostrom. 1988: "Dimensional versus Information-Processing Approaches to Social Knowledge: the Case of Inconsistency Management," in Bar-Tal & Kruglanski (1988).

Dibble, Vernon. 1964: "Four Types of Inference from Documents to Events," *History and Theory* 3:203–221.

Dietrich, Eric. 1988: "Rationality, the Frame Problem, and the Internal Manual Fallacy," Working Paper of the Computing Research Laboratory, Las Cruces: New Mexico State University.

Dietrich, Eric. 1990: "Computationalism," Social Epistemology 4:135–154.

Donovan, Arthur, Rachel Laudan, and Larry Laudan (eds.) 1988: Scrutinizing Science, Dordrecht: Kluwer.

Douglas, Mary. 1966: Purity and Danger, London: Routledge & Kegan Paul.

Downes, Stephen. 1990: The Prospects for a Cognitive Science of Science, Ph.D. dissertation, Virginia Tech.

Dreyfus, Hubert, and Stuart Dreyfus. 1986: Mind Over Machine, New York: Free Press.

Duhem, Pierre. 1954: The Aim and Structure of Physical Theory, Princeton: Princeton University Press.

Dummett, Michael. 1977: Truth and Other Enigmas, London: Duckworth.

Durkheim, Emile. 1933: The Division of Labor in Society, New York: Free Press.

Edwards, Ward and Detlof von Winterfeldt. 1986: "On Cognitive Illusions and Their Implications," in Arkes & Hammond (1986).

Elgin, Catherine. 1988: "The Epistemic Efficacy of Stupidity, Synthese 74:297–311.

Elster, Jon. 1978: Logic and Society, Chichester: John Wiley & Sons.

Elster, Jon. 1979: Ulysses and the Sirens, Cambridge UK: Cambridge University Press.

Elster, Jon. 1983: Explaining Technical Change, Cambridge UK: Cambridge University Press.

Elster, Jon. 1984a: Making Sense of Marx, Cambridge UK: Cambridge University Press.

Elster, Jon. 1984b: Sour Grapes, Cambridge UK: Cambridge University Press.

Elster, Jon (ed.) 1986: Rational Choice, Oxford: Blackwell.

Ericsson, Anders and Herbert Simon. 1984: Protocol Analysis: Verbal Reports as Data, Cambridge MA: MIT Press.

Ezrahi, Yaron. 1990: The Descent of Icarus, Cambridge MA: Harvard University Press.

Faust, David. 1984: The Limits of Scientific Reasoning, Minneapolis: University of Minnesota Press.

Feldman, Shel (ed.) 1966: Cognitive Consistency, New York: Academic Press.

Feuer, Lewis. 1963. The Scientific Intellectual, New York: Basic.

Feuer, Lewis. 1974: Einstein and the Generations of Science, New York: Basic.

Feyerabend, Paul. 1975: Against Method, London: New Left Books.

Feyerabend, Paul. 1981a: Realism, Rationalism, and the Scientific Method, Cambridge UK: Cambridge University Press.

Feyerabend, Paul. 1981b: Problems of Empiricism, Cambridge UK: Cambridge University Press.

Fields, Christopher. 1987: "The Computer as Tool," Social Epistemology 1:5–25.

Fine, Arthur. 1986a: The Shaky Game: Einstein, Realism, and Quantum Theory, Chicago: University of Chicago Press.

Fine, Arthur. 1986b: "Unnatural Attitudes: Realist and Instrumentalist Attachments to Science," Mind 95:149–179.

Finocchiaro, Maurice. 1973: History of Science as Explanation, Detroit: Wayne State University Press.

Fiske, Donald, and Richard Shweder (eds.). 1986: Metatheory in Social Science, Chicago: University of Chicago Press.

Fodor, Jerry. 1968: Psychological Explanation, New York: Random House.

Fodor, Jerry. 1975: *The Language of Thought*, New York: Thomas Crowell.

Fodor, Jerry. 1981: *Representations*, Cambridge MA: MIT Press.

Fodor, Jerry. 1983: *The Modularity of Mind*, Cambridge MA: MIT Press.

Fodor, Jerry. 1987: "Modules, Frames, Fridgeons, and the Music of the Spheres," in Pylyshyn (1987).

Foucault, Michel. 1970: *The Order of Things*, New York: Random House.

Foucault, Michel. 1977: *Discipline and Punish*, New York: Random House.

Franklin, Allan. 1986: *The Neglect of Experiment*, Cambridge UK: Cambridge University Press.

Freud, Sigmund. 1961 (1930): *Civilization and Its Discontents*, New York: Norton.

Friedman, Michael. 1983: *Foundations of Space-Time Theories*, Princeton: Princeton University Press.

Fuller, Steve. 1983: "In Search of the Science of History: the Case of Wilhelm Dilthey and Experimental Psychology," paper delivered at the philosophy department colloquium, SUNY at Stony Brook.

Fuller, Steve. 1985: "Is There a Language-Game That Even the Deconstructionist Can Play?" *Philosophy and Literature* 9:104–109.

Fuller, Steve. 1986: "Review of Apel's *Understanding and Explanation*," *Philosophy of Science* (March).

Fuller, Steve. 1987a: "Sophist vs. Skeptic: Two Paradigms of Intentional Transaction," in Otto & Tuedio (eds.), pp. 199–208.

Fuller, Steve. 1987b: "Towards Objectivism and Relativism," *Social Epistemology* 1:351–362.

Fuller, Steve. 1987c: "Deconstruction: Displacement or Elimination?" *Publication of the Society for Literature and Science* vol. 2, no. 3.

Fuller, Steve. 1988a: "Playing Without a Full Deck: Scientific Realism and the Cognitive Limits of Legal Theory," *The Yale Law Journal* 97:549–580.

Fuller, Steve. 1988b: *Social Epistemology*, Bloomington: Indiana University Press.

Fuller, Steve. 1991a: "Is History and Philosophy of Science Withering on the Vine?" *Philosophy of the Social Sciences* 21:149–174.

Fuller, Steve. 1991b: "Studying the Proprietary Grounds of Knowledge," *Journal of Social Behavior and Personality* 6(6):105–128.

Fuller, Steve. 1992a: *Philosophy, Rhetoric, and the End of Knowledge: The Coming of Science and Technology Studies*, Madison: University of Wisconsin Press.

Fuller, Steve. 1992b: "Social Epistemology and the Research Agenda of Science Studies," in Pickering (1992).

Fuller, Steve. 1992c: "Knowledge as Product and Property," in N. Stehr and R. Ericson (eds.), *The Culture and Power of Knowledge*, Berlin: Walter de Gruyter.

Fuller, Steve and David Gorman. 1987: "Burning Libraries: Cultural Creation and the Problem of Historical Consciousness," *Annals of Scholarship* 4(3):105–122.

Fuller, Steve, Marc De Mey, Terry Shinn, and Steve Woolgar (eds.). 1989: *The Cognitive Turn: Sociological and Psychological Perspectives on Science* (Sociology of the Sciences Yearbook), Dordrecht: D. Reidel.

Galison, Peter. 1987: *How Experiments End*, Chicago: University of Chicago Press.

Galison, Peter. 1993: *Image and Logic*, Chicago: University of Chicago Press.

Gardner, Howard. 1987 (1985): *The Mind's New Science*, 2nd ed., New York: Basic Books.

Garfinkel, Harold. 1963: "A Conception of, and Experiments with, 'Trust' as a Condition of Stable Concerted Actions," in O.J. Harvey (ed.), *Motivation and Social Interaction*, New York: Ronald Press.

Gaukroger, Stephen. 1975: *Explanatory Structures*, Sussex: Harvester.

Geertz, Clifford. 1973: *The Interpretation of Cultures*, New York: Harper & Row.

Gergen, Kenneth. 1983: *Towards a Transformation of Social Knowledge*, New York: Springer-Verlag.

Gergen, Kenneth. 1985: "The Social Constructionist Movement in Modern Psychology," *American Psychologist* 40:266–275.

Gholson, Barry and Arthur Houts. 1989: "Towards a Cognitive Psychology of Science," *Social Epistemology* 3:107–127.

Gholson, Barry, Arthur Houts, William Shadish, and Robert Neimeyer (eds.). 1989: *The Psychology of Science: Contributions to Metascience*, Cambridge UK: Cambridge University Press.

Gibson, James. 1979: *The Ecological Approach to Visual Perception*, Boston: Houghton Mifflin.

Giddens, Anthony. 1984: *The Constitution of Society*, Berkeley: University of California Press.

Giere, Ronald. 1988: *Explaining Science*, Chicago: University of Chicago Press.

Giere, Ronald. 1989: "The Units of Analysis in Science Studies," in Fuller et al. (1989).

Giere, Ronald (ed.) 1992: *Cognitive Models of Science*, Minneapolis: University of Minnesota Press.

Gieryn, Thomas. 1983: "Boundary-Work and the Demarcation of Science from Non-Science: Strains and Interests in the Professional Ideologies of Scientists," *American Sociological Review* 48:781–795.

Gilbert, Nigel and Michael Mulkay. 1984: *Opening Pandora's Box*, Cambridge: Cambridge University Press.

Glymour, Clark. 1980: *Theory and Evidence*, Princeton: Princeton University Press.

Glymour, Clark. 1987: "Android Epistemology and the Frame Problem," in Pylyshyn (1987).

Goldman, Alvin. 1986: *Epistemology and Cognition*, Cambridge MA: Harvard University Press.

Goldman, Alvin. 1991: *Liaisons*, Cambridge MA: MIT Press.

Goodin, Robert. 1980: *Manipulatory Politics*, Chicago: University of Chicago Press.

Goodin, Robert. 1982: *Political Theory and Public Policy*, Chicago: University of Chicago Press.

Gooding, David. 1985: "In Nature's School: Faraday as an Experimentalist," in D. Gooding and F. James (eds.), *Faraday Rediscovered: Essays on the Life and Work of Michael Faraday*, New York: Stockton Press.

Gorman, Michael. 1989: "Error, Falsification, and Scientific Inference," in Fuller et al. (1989).

Gorman, Michael and Bernard Carlson. 1989: "Can Experiments Be Used to Study Experimental Science?" *Social Epistemology* 3:89–106.

Gorman, Michael, Margaret Gorman, and R.M. Latta. 1984: "How Disconfirmatory, Confirmatory, and Combined Strategies Affect Group Problem-Solving," *British Journal of Psychology* 75:65–79.

Gorman, Michael, A. Stafford, and Margaret Gorman. 1987: "Disconfirmation and

Dual Hypotheses on a More Difficult version of Wason's 2–4–6 Task," *Quarterly Journal of Experimental Psychology* 39A:1–28.

Graumann, Carl. 1988. "From Knowledge to Cognition," in Bar-Tal & Kruglanski (1988).

Greene, Judith. 1972: *Psycholinguistics*, Baltimore: Penguin.

Greenwood, John. 1989: *Explanation and Experiment in Social Psychological Science*, New York: Springer-Verlag.

Grice, Paul. 1957: "Meaning," *Philosophical Review* 66:377–388.

Grice, Paul. 1975: "Logic and Conversation," in P. Cole & J.L. Morgan (eds.), *Speech Acts*, New York: Academic Press.

Gruber, Howard. 1981: *Darwin on Man*, Chicago: University of Chicago Press.

Gutting, Gary (ed.) 1979: *Paradigms and Revolutions*, South Bend: University of Notre Dame Press.

Hacking, Ian. (ed.) 1981a: *Scientific Revolutions*, Oxford: Oxford University Press.

Hacking, Ian. 1981b: "Lakatos' Philosophy of Science," in Hacking (1981a).

Hacking, Ian. 1983: *Representing and Intervening*, Cambridge UK: Cambridge University Press.

Hacking, Ian. 1984: "Five Parables," in Schneewind, Rorty & Skinner (1984).

Hamowy, Ronald. 1987: *The Scottish Enlightenment and the Theory of Spontaneous Order*, Carbondale: Southern Illinois University Press.

Hanson, Norwood Russell. 1958: *Patterns of Discovery*, Cambridge UK: Cambridge University Press.

Haraway, Donna. 1989: *Primate Visions*, London: Routledge.

Haraway, Donna. 1991: *Simians, Cyborgs, Women*. London: Routledge.

Hardin, Russell. 1982: *Collective Action*, Baltimore: Johns Hopkins University Press.

Harding, Sandra. 1986: *The Science Question in Feminism*, Ithaca: Cornell University Press.

Harding, Sandra. 1991: *Whose Science? Whose Knowledge?* Ithaca: Cornell University Press.

Harding, Sandra (ed.) 1987: *Feminism and Methodology*, Bloomington: Indiana University Press.

Harding, Sandra and Jean O'Barr (eds.) 1987: *Sex and Scientific Inquiry*, Chicago: University of Chicago Press.

Harre, Rom. 1970: *The Principles of Scientific Thinking*, Chicago: University of Chicago Press.

Harre, Rom. 1972: *The Philosophies of Science*, Oxford: Oxford University Press.

Harre, Rom. 1979: *Social Being*, Oxford: Blackwell.

Harre, Rom. 1984: *Personal Being*, Oxford: Blackwell.

Harre, Rom. 1986: *Varieties of Realism*, Oxford: Blackwell.

Harre, Rom and Paul Secord: 1982: *The Explanation of Social Behavior*, Oxford: Blackwell.

Harris, Marvin. 1963: *The Nature of Cultural Things*, New York: Random House.

Harris, Marvin. 1968: *The Rise of Anthropological Theory*, New York: Thomas Crowell.

Harris, Marvin. 1974. *Cows, Pigs, and Witches*, New York: Random House.

Haugeland, John. 1984: *Artificial Intelligence: The Very Idea*, Cambridge MA: MIT Press.

Hawley, Amos. 1950: *Human Ecology*, New York: The Ronald Press.

Hayes, Patrick. 1987: "What the Frame Problem Is and Isn't," in Pylyshyn (1987).

Heelan, Patrick. 1983: *Space-Perception and the Philosophy of Science*, Berkeley: University of California Press.

Hesse, Mary. 1966: *Models and Analogies in Science*, South Bend: University of Notre Dame Press.

Hirsch, E. D. 1976: *The Aims of Interpretation*, New Haven: Yale University Press.

Hirsch, E. D. 1977: *The Philosophy of Composition*, Chicago: University of Chicago Press.

Hirst, Paul and Penny Woolley. 1981: *Social Relations and Human Attributes*, London: Tavistock.

Hofstadter, Richard and Walter Metzger. 1955: *The Development of Academic Freedom in the United States*. New York: Columbia University Press.

Hogarth, Robin. 1986: "Beyond Discrete Biases: Functional and Dysfunctional Aspects of Judgmental Heuristics," in Arkes & Hammond (1986).

Holland, John, Keith Holyoak, Richard Nisbett, and Paul Thagard. 1986: *Induction*, Cambridge MA: MIT Press.

Hollis, Martin and Steven Lukes (eds.) 1982: *Rationality and Relativism*, Cambridge MA: MIT Press. Holmes, Frederick Lawrence. 1984: *Lavoisier and the Chemistry of Life*, Madison: University of Wisconsin Press.

Holton, Gerald. 1978: *The Scientific Imagination*, Cambridge UK: Cambridge University Press.

Holub, Robert. 1984: *Reception Theory*, London: Methuen.

Hooker, Clifford. 1987: *A Realistic Theory of Science*, Albany: SUNY Press.

Horwich, Paul. 1986: *Asymmetries in Time*, Cambridge MA: MIT Press.

Houts, Arthur and Barry Gholson. 1989: "Brownian Notions," *Social Epistemology* 3:139–146.

Hovland, Carl and Irving Janis, Harold Kelley. 1965 (1953): *Communication and Persuasion*, New Haven: Yale University Press.

Hoy, David. 1978: *The Critical Circle*, Berkeley: University of California Press.

Hull, David. 1974: *The Philosophy of Biological Science*, Englewood-Cliffs: Prentice-Hall.

Hull, David. 1983: "Exemplars and Scientific Change," in P. Asquith & T. Nickles (eds.), *PSA 1982*, vol. 2, pp. 479–503.

Hull, David. 1988: *Science as a Process*, Chicago: University of Chicago Press.

Husserl, Edmund. 1970: *Crisis in European Sciences and Transcendental Phenomenology*, Evanston: Northwestern University Press.

Ihde, Don. 1987: *Technology and the Lifeworld*, Bloomington: Indiana University Press.

Ihde, Don. 1991: *Instrumental Realism*, Bloomington: Indiana University Press.

Jacobs, Margaret. 1976: *The Newtonians and the English Revolution: 1689–1720*, Ithaca: Cornell University Press.

Jacobs, R. C. and D. T. Campbell. 1961: "The Perpetuation of an Arbitrary Tradition Through Several Generations of a Laboratory Microculture," *Journal of Abnormal and Social Psychology* 62:649–658.

Janlert, Lars-Erik. 1987: "Modeling Change—The Frame Problem," in Pylyshyn (1987).

Jansen, Sue Curry. 1988: *Censorship*, Oxford: Oxford University Press.

Jardine, Nicholas. 1991: *The Scenes of Inquiry*, Oxford: Oxford University Press.

Jonsen, Albert and Stephen Toulmin. 1988: *The Abuse of Casuistry*, Berkeley: University of California Press.

Jungermann, Helmut. 1986: "The Two Camps on Rationality," in Arkes & Hammond (1986).

Keller, Evelyn Fox. 1985: *Reflections on Gender and Science*, New Haven: Yale University Press.

Kitcher, Philip. 1982: *Abusing Science*, Cambridge MA: MIT Press.

Kitcher, Philip. 1985: *Vaulting Ambition*, Cambridge MA: MIT Press.

Knorr-Cetina, Karin. 1980: *The Manufacture of Knowledge*, Oxford: Pergamon.

Knorr-Cetina, Karin and Aaron Cicourel (eds.). 1981: *Advances in Social Theory*, London: Routledge & Kegan Paul.

Koestler, Arthur. 1959: *The Sleepwalkers*, London: Hutchinson.

Kornblith, Hilary (ed.) 1985: *Naturalizing Epistemology*, Cambridge MA: MIT Press.

Kornblith, Hilary. 1987: "Some Social Features of Cognition," *Synthese* 73:27–42.

Kornblith, Hilary. 1988: "How Internal Can You Get?" *Synthese* 74:313–327.

Koyre, Alexandre. 1978: *Galilean Studies*, Sussex: Harvester.

Kruglanski, Arie. 1989: *Lay Epistemics and Human Knowledge*, New York: Plenum Press.

Kuhn, Thomas. 1970a (1962): *The Structure of Scientific Revolutions*, 2nd ed., Chicago: University of Chicago Press.

Kuhn, Thomas. 1970b: "Reflections on My Critics," in Lakatos & Musgrave (1970).

Kuhn, Thomas. 1977: *The Essential Tension*, Chicago: University of Chicago Press.

Kuhn, Thomas. 1981: "A Function for Thought-Experiments," in Hacking (1981a), pp. 6–27.

Kyburg, Henry. 1983: "Rational Belief," *Behavioral and Brain Sciences* 2:535–581.

Kyburg, Henry. 1987: "The Hobgoblin," *Monist* 70:141–151.

Lakatos, Imre. 1970: "Falsification and the Methodology of Scientific Research Programs," in Lakatos & Musgrave (1970).

Lakatos, Imre. 1981: "History of Science and Its Rational Reconstructions," in Hacking (1981a).

Lakatos, Imre and Alan Musgrave (eds.). 1970: *Criticism and the Growth of Knowledge*, Cambridge UK: Cambridge University Press.

Langley, Pat, Herbert Simon, Gary Bradshaw, and Jan Zytkow. 1987: *Scientific Discovery: Computational Explorations of the Creative Process*, Cambridge MA: MIT Press.

Latour, Bruno. 1986: "Visualization and Cognition," in A. Pickering (ed.), *Knowledge and Society*, vol. 6, Greenwich CT: JAI Press, pp. 1–40.

Latour, Bruno. 1987a: *Science in Action*, Milton Keynes: Open University Press.

Latour, Bruno. 1987b: "Clothing the Naked Truth," in H. Lawson and L. Appignanesi (eds.), *Dismantling Truth: Objectivity and Science*, London: Sage.

Latour, Bruno. 1988: "The Politics of Explanation: An Alternative," in Woolgar (1988a).

Latour, Bruno. 1989: *The Pasteurization of France*, Cambridge MA: Harvard University Press.

Latour, Bruno and Steve Woolgar. 1979: *Laboratory Life*, London: Sage.

Laudan, Larry. 1977: *Progress and Its Problems*, Berkeley: University of California Press.

Laudan, Larry. 1981: "A Problem-Solving Approach to Scientific Progress," in Hacking (1981a).

Laudan, Larry. 1982: *Science and Hypothesis*, Dordrecht: Reidel.

Laudan, Larry. 1983: "The Demise of the Demarcation Criterion," in Virginia Tech

Working Papers on the Demarcation of Science and Non-Science, Blacksburg: Virginia Tech.

Laudan, Larry. 1984: *Science and Values*, Berkeley: University of California Press.

Laudan, Larry. 1986: "Some Problems Facing Intuitionist Meta-Methodologies," *Synthese* 67:115–129.

Laudan, Larry. 1987: "Progress or Rationality? The Prospects for Normative Naturalism, *American Philosophical Quarterly* 24:19–31.

Laudan, Larry. 1990: *Science and Relativism*, Chicago: University of Chicago Press.

Laudan, Larry and Arthur Donovan, Rachel Laudan, Peter Barker, Harold Brown, Jarrett Leplin, Paul Thagard, Steve Wykstra. 1986: "Testing Theories of Scientific Change," *Synthese* 69:141–223.

Lea, Stephen and Roger Tarpy, Paul Welby. 1987: *The Individual in the Economy*, Cambridge UK: Cambridge University Press.

Lehrer, Keith and Carl Wagner. 1986: *Rational Consensus in Science and Society*, Dordrecht: D. Reidel.

Leplin, Jarrett (ed.) 1984a: *Scientific Realism*, Berkeley: University of California Press.

Leplin, Jarrett. 1984b: "Truth and Scientific Progress," in Leplin (1984a).

Levi, Isaac. 1987: "The Demons of Decision," *Monist* 70:193–211.

Levins, Richard and Richard Lewontin. 1985: *The Dialectical Biologist*, Cambridge MA: Harvard University Press.

Levi-Strauss, Claude. 1964: *Structural Anthropology*, New York: Harper & Row.

Levy-Bruhl, Henri. 1978: *Notebooks on the Primitive Mentality*, New York: Harper & Row.

Lindenfeld, David. 1980: *Towards a Transformation of Positivism: 1880–1920*, Berkeley: University of California Press.

Lindholm, Lynn. 1981: "Is Realistic History of Science Possible?" in J. Aggasi & R. Cohen (eds.) *Scientific Philosophy Today*, Dordrecht: D. Reidel.

Lipsey, R.G. and K. Lancaster. 1956: "The General Theory of the Second Best," *Review of Economic Studies* 24:11–32.

Longino, Helen. 1990: *Science as Social Knowledge*, Princeton: Princeton University Press.

Lowe, Adolph. 1965: *On Economic Knowledge*, New York: Harper & Row.

Luhmann, Niklas. 1979: *The Differentiation of Society*, New York: Columbia University Press.

Lynch, Michael and Steve Woolgar (eds.) 1990: *Representation in Scientific Practice*, Cambridge MA: MIT Press.

Lyons, William. 1986: *The Disappearance of Introspection*, Cambridge MA: MIT Press.

MacDonald, Graham and Philip Pettit. 1981: *Semantics and Social Science*, London: Routledge & Kegan Paul.

Mach, Ernst. 1943: *Popular Scientific Lectures*, La Salle: Open Court.

Machlup, Fritz. 1978: *Methodology of Economics and Other Social Sciences*, New York: Academic Press.

MacIntyre, Alasdair. 1984: *After Virtue*, South Bend: Notre Dame University Press.

MacIver, Robert. 1947: *The Web of Government*, New York: Macmillan.

Manicas, Peter. 1986: *A History and Philosophy of the Social Sciences*, Oxford: Blackwell.

Manier, Edward. 1986: "Social Dimensions of the Mind/Body Problem: Turbulence in the Flow of Scientific Information," *Science and Technology Studies*, 4(3/4):16–28.

Mannheim, Karl. 1936 (1929): *Ideology and Utopia*, London: Routledge & Kegan Paul.

Manuel, Frank. 1969: *A Portrait of Isaac Newton*, Cambridge MA: Harvard University Press.

March, James. 1978: "Bounded Rationality, Ambiguity, and the Engineering of Choice," *Bell Journal of Economics* 9:587–608.

Margalit, Avishai. 1986: "The Past of an Illusion: Comment on Tversky," in E. Ullmann-Margalit (ed.), *The Kaleidoscope of Science*, Dordrecht: D. Reidel.

Maturana, Humberto and Francisco Varela. 1980. *Autopoiesis and Cognition*, Dordrecht: D. Reidel.

McCloskey, Donald. 1985: *The Rhetoric of Economics*, Madison: University of Wisconsin Press.

McDermott, Drew. 1987: "We've Been Framed," in Pylyshyn (1987).

McLuhan, Marshall. 1965: *The Gutenberg Galaxy*, Toronto: University of Toronto Press.

McMullin, Ernan. 1984: "The Rational and the Social in the History of Science," in Brown (1984).

Mead, George Herbert. 1934: *Mind, Self, and Society*, Chicago: University of Chicago Press.

Meehl, Paul. 1984: "Foreword" to Faust (1984).

Meja, Volker and Nico Stehr. 1988: "Social Science, Epistemology, and the Problem of Relativism," *Social Epistemology* 2:263–271.

Merton, Robert. 1968 (1949): *Social Theory and Social Structure*, New York: Free Press.

Merton, Robert. 1976: *Sociological Ambivalence*, New York: Free Press.

Merton, Robert. 1977: *The Sociology of Science*, Chicago: University of Chicago Press.

Mill, John Stuart. 1843: *A System of Logic*, 2 vols., London.

Minsky, Marvin. 1986: *The Society of Mind*, New York: Simon & Schuster.

Mirowski, Philip. 1989: *More Heat Than Light*, Cambridge UK: Cambridge University Press.

Mitroff, Ian. 1974: *The Subjective Side of Science*, Amsterdam: Elsevier.

Morawski, Jill (ed.) 1988: *The Rise of Experimentation in American Psychology*, New Haven: Yale University Press.

Mulkay, Michael. 1990: *The Sociology of Science*, Bloomington: Indiana University Press.

Mynatt, Clifford, Michael Doherty, and Ryan Tweney. 1978: "Consequences of Confirmation and Disconfirmation in a Simulated Research Environment," *Quarterly Journal of Experimental Psychology* 30:395–406.

Nagel, Ernest. 1960: *The Structure of Science*, New York: Free Press.

Nagel, Thomas, 1979: *Mortal Questions*, Oxford: Oxford University Press.

Nagel, Thomas. 1986: *The View from Nowhere*, Oxford: Oxford University Press.

Neimeyer, Robert and William Shadish. 1987: "Optimizing Scientific Validity," *Knowledge* 8:463–485.

Nersessian, Nancy. 1984: *Faraday to Einstein: Constructing Meaning in Scientific Theories*, Dordrecht: Martinus Nijhoff.

Nersessian, Nancy (ed.) 1987: *The Process of Science*, Dordrecht: Martinus Nijhoff.

Naess, Arne. 1970: *Skepticism*, London: Routledge & Kegan Paul.

Nelson, John, Allan Megill, Donald McCloskey (eds.) 1987: *The Rhetoric of the Human Sciences*, Madison: University of Wisconsin Press.

Nickles, Thomas (ed.) 1980a: *Scientific Discovery, Logic and Rationality*, 2 vols., Dordrecht: Reidel.

Nickles, Thomas. 1980b: "Can Scientific Constraints Be Rationally Violated?" in Nickles (1980a).

Nickles, Thomas. 1985: "Beyond Divorce: Current Status of the Discovery Debate," *Philosophy of Science* 52:177–206.

Nickles, Thomas. 1986: "Remarks on the Use of History as Evidence," *Synthese* 69:253–266.

Nickles, Thomas. 1987: "'Twixt Method and Madness," in Nersessian (1987).

Nisbett, Richard and Timothy Wilson: 1977, "Telling More Than We Can Know," *Psychological Review* 84:231–259.

Noelle-Neumann, Elisabeth. 1984: *The Spiral of Silence*, Chicago: University of Chicago Press.

Oldroyd, David. 1986: *The Arch of Knowledge*, London: Methuen.

Ong, Walter. 1958: *Ramus, Method, and the Decay of Dialogue*, Cambridge MA: Harvard University Press.

Ormiston, Gayle and Raphael Sassower. 1989: *Narrative Experiments*, Minneapolis: University of Minnesota Press.

Ortony, Andrew (ed.) 1979: *Metaphor and Thought*, Cambridge UK: Cambridge University Press.

Otto, Herbert and James Tuedio (eds.) 1987: *Perspectives on Mind*, Dordrecht: D. Reidel.

Outhwaite, William. 1983: *Concept Formation in the Social Sciences*, London: Routledge & Kegan Paul.

Parfit, Derek. 1984. *Reasons and Persons*, Oxford: Oxford University Press.

Park, Robert and Ernest Burgess. 1921: *Introduction to the Science of Sociology*, Chicago: University of Chicago Press.

Parsons, Talcott. 1951: *The Social System*, New York: Free Press.

Passmore, John. 1966: *A Hundred Years of Philosophy*, Harmondsworth: Penguin.

Peacocke, Christopher. 1988: "The Limits of Intelligibility," *Philosophical Review* 97:463–496.

Peirce, Charles Sanders. 1964: *The Essential Writings*, New York: Dover.

Pettit, Philip and John MacDowell (eds.) 1986: *Subject, Thought, and Context*, Oxford: Oxford University Press.

Pickering, Andrew. 1984: *Constructing Quarks*, Chicago: University of Chicago Press.

Pickering, Andrew (ed.) 1992: *Science as Practice and Culture*, Chicago: University of Chicago Press.

Pitt, Joseph. 1988: "Progressive Science," *Social Epistemology* 2:341–344.

Polanyi, Michael. 1957: *Personal Knowledge*, Chicago: University of Chicago Press.

Pollock, John. 1986: *Contemporary Theories of Knowledge*, London: Hutchinson.

Popper, Karl. 1957: *The Poverty of Historicism*, New York: Harper & Row.

Popper, Karl. 1959 (1934): *The Logic of Scientific Discovery*, New York: Harper & Row.

Popper, Karl. 1963: *Conjectures and Refutations*, New York: Harper & Row.

Popper, Karl. 1972: *Objective Knowledge*, Oxford: Oxford University Press.

Popper, Karl. 1981: "The Rationality of Scientific Revolutions," in Hacking (1981a).

Popper, Karl. 1984: *Postscript to the Logic of Scientific Discovery*, La Salle: Open Court.

Porter, Theodore. 1986: *The Rise of Statistical Thinking: 1820–1900*, Princeton: Princeton University Press.

Prendergast, Christopher. 1986: "Alfred Schutz and the Austrian School of Economics," *American Journal of Sociology* 92:1–26.

Prigogine, Ilya and Isabelle Stengers. 1984: *Order Out of Chaos*, New York: Bantam Books.

Proctor, Robert. 1991: *Value-Free Science?*, Cambridge MA: Harvard University Press.

Putnam, Hilary. 1973: "Explanation and Reference," in G. Pearce and P. Maynard (eds.), *Conceptual Change*, Dordrecht: D. Reidel.

Putnam, Hilary. 1975: *Mind, Language, and Reality*, Cambridge UK: Cambridge University Press.

Putnam, Hilary. 1978: *Meaning and the Moral Sciences*, London: Routledge & Kegan Paul.

Putnam, Hilary. 1983: *Realism and Reason*, Cambridge UK: Cambridge University Press.

Putnam, Hilary. 1984: "What Is Realism?" in Leplin (ed.) (1984a).

Putnam, Hilary. 1987: *The Many Faces of Realism*, La Salle: Open Court.

Pylyshyn, Zenon (ed.) 1970: *Perspectives on the Computer Revolution*, New York: Prentice-Hall.

Pylyshyn, Zenon. 1973: "What the Mind's Eye Tells the Mind's Brain," *Psychological Bulletin* 8:1–14.

Pylyshyn, Zenon. 1979: "Metaphorical Imprecision and the 'Top-Down' Research Strategy," in Ortony (1979).

Pylyshyn, Zenon. 1984: *Computation and Cognition*, Cambridge MA: MIT Press.

Pylyshyn, Zenon (ed.) 1987: *The Robot's Dilemma*, Norwood NJ: Ablex.

Quine, W.V.O. 1953: *From a Logical Point of View*, New York: Harper & Row.

Quine, W.V.O. 1960: *Word and Object*, Cambridge MA: MIT Press.

Quine, W.V.O. 1969: "Epistemology Naturalized," in *Ontological Relativity*, New York: Columbia University Press.

Rachlin, Howard. 1980: "Economics and Behavioral Psychology," in J.E.R. Staddon (ed.), *Limits to Action: The Allocation to Individual Behavior*, New York: Academic Press.

Rachlin, Howard, A. Logue, J. Gibbon, and M. Frankel. 1986: "Cognition and Behavior in Studies of Choice," *Psychological Review* 93:33–45.

Rapoport, Anatol. 1980: "Various Meanings of 'Rational Political Decisions'," in L. Lewin & E. Vedung (eds.), *Politics as Rational Action*, Dordrect: Reidel, pp. 39–59.

Rawls, John. 1955: "Two Concepts of Rules," *Philosophical Review* 64.

Redner, Harry. 1986: *The Ends of Philosophy*, London: Croom Helm.

Redner, Harry. 1987: *The Ends of Science*, Boulder: Westview.

Reichenbach, Hans. 1938: *Experience and Prediction*, Chicago: University of Chicago Press.

Rescher, Nicholas. 1977: *Dialectics*, Albany: SUNY Press.

Rescher, Nicholas. 1979: *Scientific Progress*, Oxford: Blackwell.

Ruben, David-Hillel. 1986: *The Metaphysics of the Social World*, London: Routledge & Kegan Paul.

Rich, Robert. 1983: "Knowledge Synthesis and Problem Solving," in S. Ward & L. Reed (eds.), *Knowledge Structure and Use*, Philadelphia: Temple University Press.

Richards, Robert. 1987: *Darwin and the Emergence of Evolutionary Theories of Mind and Behavior*, Chicago: University of Chicago Press.

Rickert, Heinrich. 1986 (1902): *The Limits of Concept Formation in Natural Science*, Cambridge UK: Cambridge University Press.

Ricoeur, Paul. 1970: *Freud and Philosophy*, New Haven: Yale University Press.

Rorty, Richard. 1979: *Philosophy and the Mirror of Nature*, Princeton: Princeton University Press.

Rorty, Richard. 1982: *Consequences of Pragmatism*, Minneapolis: University of Minnesota Press.

Rosenberg, Alexander. 1980: *Sociobiology and the Preemption of Social Science*, Baltimore: Johns Hopkins University Press.

Rosenberg, Alexander. 1985: *The Structure of Biological Science*, Cambridge UK: Cambridge University Press.

Rosenberg, Alexander. 1989: *Philosophy of Social Science*, Boulder: Westview.

Rosenfeld, Israel. 1987: *The Invention of Memory*, New York: Basic Books.

Rosenthal, Peggy. 1984: *Words and Values*, Oxford: Oxford University Press.

Ross, Lee. 1977: "The Intuitive Psychologist and His Shortcomings," in L. Berkowitz (ed.), *Advances in Experimental Social Psychology*, New York: Academic Press.

Ross, Michael and Cathy McFarland. 1988: "Constructing the Past: Biases in Personal Memories," in Bar-Tal & Kruglanski (1988).

Roth, Paul. 1987: *Meaning and Method in the Social Sciences*, Ithaca: Cornell University Press.

Rouse, Joseph. 1987: *Knowledge and Power: Toward a Political Philosophy of Science*, Ithaca: Cornell University Press.

Ruben, David-Hillel, 1986: *The Metaphysics of the Social World*, London: Routledge.

Runciman, W.G. 1989: *A Treatise on Social Theory, Vol. II: Substantive Social Theory*, Cambridge UK: Cambridge University Press.

Ruse, Michael. 1986: *Taking Darwin Seriously*, Oxford: Blackwell.

Sarkar, Husain. 1983: *A Theory of Method*, Berkeley: University of California Press.

Schaefer, Wolf. (ed.) 1984: *Finalization in Science*, Dordrecht: Reidel.

Schlick, Moritz. 1964: "The Future of Philosophy," in D. Bronstein et al. (eds.), *The Basic Problems of Philosophy*, 3rd ed., Englewood Cliffs: Prentice-Hall.

Schmaus, Warren. 1988: "Reply to Meja & Stehr," Social Epistemology, 2:273–274.

Schneewind, Jerome, Richard Rorty and Quentin Skinner (eds.) 1984: *Philosophy in History*, Cambridge UK: Cambridge University Press.

Schneewind, Jerome. 1987: "The Use of Autonomy in Ethical Theory," in T. Heller, M. Sosna, D. Wellerby (eds.), *Reconstructing Individualism*, Palo Alto: Stanford University Press.

Schwartz, Stephen (ed.) 1977: *Naming, Necessity, and Natural Kinds*, Ithaca: Cornell University Press.

Searle, John. 1983: *Intentionality*, Cambridge UK: Cambridge University Press.

Searle, John. 1984: *Minds, Brains, and Science*, Cambridge MA: Harvard University Press.

Segall, Marshall, Donald Campbell and Melville Herskovitz. 1966: *The Influence of Culture on Visual Perception*, Indianapolis: Bobbs-Merrill.

Shadish, William. 1989: "From Program Evaluation to Science Evaluation," *Social Epistemology* 3:189–204.

Shadish, William, Thomas Cook, and Laura Leviton. 1991: *Foundations of Program Evaluation*, Newbury Park CA: Sage.

Shadish, William and Steve Fuller (eds.) 1992: *The Social Psychology of Science*, New York: Guilford Press.

Shadish, William and Robert Neimeyer. 1989: "Contributions of Psychology to an

Integrative Science Studies: The Shape of Things to Come," in Fuller et al. (1989).

Shapere, Dudley. 1984 (1960–82): *Reason and the Search for Knowledge*, Dordrecht: D. Reidel.

Shapere, Dudley. 1987: "Method in the Philosophy of Science and Epistemology: How to Inquire About Inquiry and Knowledge," in Nersessian (1987).

Shapin, Steven and Simon Schaffer. 1985: *Leviathan and the Air-Pump*, Princeton: Princeton University Press.

Shotter, John. 1984: *Social Accountability and Selfhood*, Oxford: Blackwell.

Shrager, Jeff and Pat Langley (eds.) 1990: *Computational Models of Scientific Discovery and Theory Formation*, San Mateo: Morgan Kauffman.

Shweder, Richard. 1987: "Comments on Plott, and on Kahneman, Knetsch, and Thaler," in R. Hogarth & M. Reder (eds.) *Rational Choice*, Chicago: University of Chicago Press.

Shweder, Richard and Edmund Bourne. 1984: "Does the Concept of Person Vary Cross-Culturally?" in Shweder & LeVine (1984).

Shweder, Richard and Robert LeVine (eds.) 1984: *Culture Theory*, Cambridge UK: Cambridge University Press.

Simmel, Georg. 1950: *The Sociology of Georg Simmel*, New York: Free Press.

Simon, Herbert. 1981 (1969): *The Sciences of the Artificial*, 2nd ed., Cambridge MA: MIT Press.

Simonton, Dean. 1988: *Genius in Science: A Psychology of Science*, Cambridge UK: Cambridge University Press.

Skinner, B.F. 1954: *Science and Human Behavior*, New York: Free Press.

Skinner, B.F. 1957: *Verbal Behavior*, New York: Appleton-Century Crofts.

Skinner, Quentin. 1969: "Meaning and Understanding in the History of Ideas," *History and Theory* 8:3–53.

Slezak, Peter. 1989: "Scientific Discovery by Computer as Empirical Refutation of the Strong Programme," *Social Studies of Science* 19:563–600.

Small, Henry. 1986: "The Synthesis of Specialty Narratives from Co-Citation Clusters," *Journal of the American Society for Information Science* 37:97–110.

Sober, Elliott. 1984: *The Nature of Selection*, Cambridge MA: MIT Press.

Sober, Elliott and Richard Lewontin. 1984: "Artifact, Cause, and Genic Selection," in Brandon & Burian (1984).

Soper, Kate. 1986: *Humanism and Anti-Humanism*, La Salle: Open Court.

Sorokin, Pitirim. 1928: *Contemporary Sociological Theories*, New York: Harper & Row.

Sowell, Thomas. 1987: *A Conflict of Visions*, New York: Morrow.

Stich, Stephen. 1982: "Dennett on Intentional Systems," in Biro & Shahan (1982).

Stich, Stephen. 1983: *From Folk Psychology to Cognitive Science*, Cambridge: MIT Press.

Stich, Stephen. 1985: "Could Man Be an Irrational Animal?" in Kornblith (1985).

Stich, Stephen. 1986: "Are Belief Predicates Systematically Ambiguous?" in R. Bogdan (ed.) *Belief: Form Content, and Function*, Oxford: Oxford University Press.

Stich, Stephen. 1988: "Reflective Equilibrium, Analytic Epistemology, and the Problem of Cognitive Diversity," *Synthese* 74:391–413.

Stich, Stephen. 1990: *The Fragmentation of Reason*, Cambridge MA: MIT Press.

Stich, Stephen and Richard Nisbett. 1980: "Justification and the Psychology of Human Reasoning," *Philosophy of Science* 47:188–202.

Stocking, George. 1968: *Race, Culture, and Evolution*, New York: Free Press.

Stout, Jeffrey. 1984: *The Flight From Authority*, South Bend: University of Notre Dame Press.

Strauss, Leo. 1958: *Thoughts on Machiavelli*, Chicago: University of Chicago Press.

Strawson, Peter. 1959: *Individuals*, London: Methuen.

Suppe, Frederick (ed.). 1977: *The Structure of Scientific Theories*, 2nd ed., Urbana: University of Illinois Press.

Taylor, Charles. 1982: "Rationality," in Hollis & Lukes (1982).

Taylor, Charles. 1985: *Human Agency and Language*, Cambridge UK: Cambridge University Press.

Thagard, Paul. 1988: *Computational Philosophy of Science*, Cambridge MA: MIT Press.

Thiem, John. 1979: "The Great Library of Alexandria Burnt: Towards the History of a Symbol," *Journal of the History of Ideas* 40:507–526.

Thompson, John. 1984: *Studies in the Theory of Ideology*, Berkeley: University of California Press.

Toulmin, Stephen. 1968: "The Complexity of Scientific Choice II: Culture, Overheads, or Tertiary Industry?" in E. Shils (ed.), *Criteria for Scientific Development*, Cambridge MA: MIT Press.

Toulmin, Stephen. 1972: *Human Understanding*, Princeton: Princeton University Press.

Turner, Stephen. 1986: "The Sociology of Science in Its Place: Comment on Shapere," *Science and Technology Studies* 4:14–17.

Tversky, Amos and Daniel Kahneman. 1986: "The Framing of Decision and the Psychology of Choice," in Elster (1986).

Tversky, Amos and Daniel Kahneman. 1987: "Can Normative and Descriptive Analysis Be Reconciled?" Working Paper of the Center for Philosophy and Public Policy. College Park: University of Maryland.

Tweney, Ryan. 1989: "A Framework for the Cognitive Psychology of Science," in Gholson et al. (1989).

Tweney, Ryan. 1990: "Five Questions for Computationalists," in Shrager and Langley (1990).

Tweney, Ryan, Michael Doherty and Clifford Mynatt (eds.) 1981: *On Scientific Thinking*, New York: Columbia University Press.

Van Fraassen, Bas. 1980: *The Scientific Image*, Oxford: Oxford University Press.

Von Hayek, Friedrich. 1985: *New Studies in Philosophy, Politics, Economics, and the History of Ideas*, Chicago: University of Chicago Press.

Von Wright, Georg. 1971: *Explanation and Understanding*, Ithaca: Cornell University Press.

Webb, Eugene, Donald Campbell, Richard Schwartz, and Lee Sechrest. 1969: *Unobtrusive Measures: Nonreactive Research in the Social Sciences*, Chicago: Rand-McNally.

Weber, Max. 1954: "Science as a Vocation," in H. Gerth & C.W. Mills (eds.), *From Max Weber*, New York: Columbia University Press.

Weber, Max. 1964 (1904–19): *The Methodology of the Social Sciences*, New York: Free Press.

Whitley, Richard. 1986: *The Social and Intellectual Organization of the Sciences*, Oxford: Oxford University Press.

Wilkes, Kathleen. 1988: *Real People*, Oxford: Oxford University Press.

Willard, Charles. 1983: *Argumentation and the Social Grounds of Knowledge*, Tuscaloosa: University of Alabama Press.

Willard, Charles. 1992: *Liberalism and the Problem of Competence*, manuscript.

Williams, Bernard. 1973: *Problems of the Self*, Cambridge UK: Cambridge University Press.

Williams, Bernard. 1981: *Moral Luck*, Cambridge UK: Cambridge University Press.

Wilson, Bryan. (ed.) 1970: *Rationality*, Oxford: Blackwell.

Wimsatt, William. 1984: "Reductionist Research Strategies and Their Biases in the Units of Selection Controversy," in Brandon & Burian (1984).

Winch, Peter. 1958: *The Idea of a Social Science*, London: Routledge & Kegan Paul.

Wittgenstein, Ludwig. 1958: *Philosophical Investigations*, Oxford: Oxford University Press.

Woodfield, Andrew (ed.) 1982: *Thoughts and Objects*, Oxford: Oxford University Press.

Woolgar, Steve. 1981: "Interests and Explanation in the Social Study of Science," *Social Studies of Science* 11:365–394.

Woolgar, Steve. 1985: "Why Not a Sociology of Machines?" *Sociology* 19:557–572.

Woolgar, Steve (ed.) 1988a: *Knowledge and Reflexivity*, London: Sage.

Woolgar, Steve. 1988b: *Science: The Very Idea*, London: Tavistock.

Wright, Crispin. 1980: *Wittgenstein and the Foundations of Mathematics*, London: Duckworth.

Wuebben, Paul, Bruce Straits, and Gary Shulman (eds.). 1974: *The Experiment as a Social Occasion*, Berkeley: Glendessary Press.

Wyer, Robert and Thomas Srull. 1988: "Understanding Social Knowledge: If Only the Data Could Speak for Themselves," in Bar-Tal & Kruglanski (1988).

Index